Sustainable Mobility & Energy

Series Editor

Helmut List, List Gmbh, Anstalt Verbrennungskraftmaschinen, Graz, Austria

Sustainable mobility is a decisive factor in reducing CO_2 emissions and combating climate change.

Electric vehicles offer an environmentally friendly alternative to conventional combustion engines, especially when the electricity required comes from renewable energy sources such as wind or solar power. Utilising these clean energies contributes significantly to reducing air pollution and reduces dependence on fossil fuels.

Fuel cells are gaining in importance alongside electromobility. This technology converts hydrogen into electrical energy and only emits water vapour, making it particularly environmentally friendly. Hydrogen can also be produced from surplus renewable energy, creating a sustainable circular economy.

Another aspect of sustainable mobility is networked, cooperative and automated mobility. By integrating modern communication technologies, vehicles can interact with each other and with the infrastructure. This enables more efficient traffic management, reduces congestion and increases road safety. Automated vehicles based on artificial intelligence and advanced sensor systems can optimise traffic flow and reduce energy consumption.

Overall, the combination of these innovative approaches offers a promising solution for the mobility requirements of the future that is both environmentally friendly and efficient. By promoting electromobility, the use of fuel cells and the development of networked, cooperative and automated mobility, we can make an important contribution to protecting our planet.

Daniel Watzenig
Editor

Recent Advances in Autonomous Vehicle Technology—Perception and Path Planning

Springer

Editor
Daniel Watzenig
Institute of Visual Computing
Graz University of Technology and CTO
Virtual Vehicle Research
Graz, Austria

ISSN 2731-9571 ISSN 2731-958X (electronic)
Sustainable Mobility & Energy
ISBN 978-3-032-01000-1 ISBN 978-3-032-01001-8 (eBook)
https://doi.org/10.1007/978-3-032-01001-8

This Springer imprint is published by the registered company Springer Nature Switzerland AG
The registered company address is: Gewerbestrasse 11, 6330 Cham, Switzerland

Series Foreword

In the year 2002, a new series was created under the title "Internal Combustion Engine—Powertrain", with a new layout and a new Scientific Advisory Board. This series shows the holistic context of the individual components of a powertrain. Several books were published. In particular, the English language editions were promoted next to the German editions.

In this series, the classification was made according to the areas of responsibility in the industry at that time.

In the years 2014 to 2020 a subsequent series named "Vehicle Propulsion" was created, also with the new advisory board. From 2002 to 2022, 21 volumes were published.

Today, the automotive industry is in a transition period with a lot of new technology and therefore with increased complexity. This was the reason to create a new series focusing on the research and development activities for today's and future vehicles, in particular also focusing on the complex powertrain system.

This book series has a similar focus as the last series, especially on the state of knowledge in the various fields in the industry, starting from basics and describing the necessary background information. Today, the system approach, next to the various specific technologies, is of high importance, and the book series also takes this approach into consideration. We are sure that some contents of this book series won't be state of the art in a few years, as the R&D activities for technologies are advancing quickly. Here, the digital version of the Springer publication is an essential tool for doing so. The book series not only covers the powertrain system with battery or fuel cell, but also with the influence and the needed aspects of controlled, automated mobility. The topic of mobility services will be addressed for the assessment of subjective and environmental aspects in each volume. In addition to the technical content, methods, processes and visions to solve the complexity of the future powertrain system by simulation are presented. The circumstances of different economies and their respective requirements are set out in concepts.

In particular, this book series will also contribute to the sustainable development goals of the United Nations.

The book series provides both students at universities and colleges as well as practitioners in the industry with a guidebook from which they can acquire the accumulated experience of the authors' knowledge.

In advance, I thank the editors and authors who have agreed to share their knowledge in this book series and will invest their efforts in writing them. I also want to thank our publisher, Springer, for their cooperation. I am very grateful to the new Scientific Advisory Board, which stood by me in both the division of the very broad subject area as well as in the selection of editors and authors. The members of the board are: Prof. Helmut Eichlseder, Graz University of Technology; Stefan Finkbeiner, Bosch; Prof. Bernhard Geringer, Technical University of Vienna; Prof. Barbara Lenz, DLR/University Berlin; Prof. Jun Li, Tsinghua University China; Prof. Christian Mohrdieck, Hyzon; Stephan Neugebauer, BMW; Dr. Steven E. Shladover, University of California/Berkeley and Mats Rosenquist, Volvo Truck. I would also like to thank J. Affenzeller for the coordination of our editorial activities.

Graz, Austria
2026

Helmut List

Preface

The field of autonomous vehicle technology has witnessed remarkable advancements in recent years, driven by the convergence of artificial intelligence, sensor technology, and sophisticated control systems. Among the most critical aspects of this evolution are perception and path planning—two fundamental pillars that enable self-driving vehicles to navigate complex environments safely and efficiently. This book, *Recent Advances in Autonomous Vehicle Technology—Perception and Path Planning*, offers a timely exploration of cutting-edge research and innovative methodologies that push the boundaries of what is possible in autonomous driving.

One of the recurring themes throughout the chapters of this book is the necessity of robust impact analysis in hybrid traffic environments, where automated and human-controlled vehicles coexist. Quantifying the impact of automated vehicles by partitioning operational design domains into specific scenarios provides a nuanced understanding essential for optimizing performance and safety. Equally critical is the resilience of deep learning systems that underpin perception in autonomous vehicles. As highlighted in this collection, the vulnerability of deep neural networks to adversarial examples remains a pressing concern. By reviewing attack and defense mechanisms developed over the past decade, the book underscores the importance of trustworthy AI in automated driving. Dynamic occupancy grid mapping (DOGM) is another notable area of innovation discussed herein. The computational demands of radar-based DOGM algorithms present challenges in real-time applications. This book introduces optimization strategies that reduce computation time and variability, offering valuable insights for developers striving for efficiency without compromising performance. Further enriching the discourse is a focus on LiDAR point cloud processing, where the integration of local geometric features into self-supervised pre-training models significantly enhances semantic segmentation capabilities. This advancement emphasizes the growing importance of sophisticated feature extraction methods in improving perception accuracy. On the planning and control front, the application of reinforcement learning (RL) in vehicle stability control systems marks a shift towards more adaptive and optimal control strategies. The use of high-fidelity simulation models demonstrates RL's potential to outperform traditional algorithms, especially in complex maneuvers like obstacle avoidance. The exploration of scaled robotic car platforms provides a practical perspective on testing automated driving algorithms. Benchmarking multiple path tracking control strategies on scaled models allows for cost-effective experimentation and reveals the strengths of model predictive and sliding mode controllers in challenging scenarios. Finally, addressing the issue of dead time in path tracking control systems reflects the ongoing pursuit of precision in real-world applications. The proposed nonlinear prediction model demonstrates how innovative compensation methods can significantly enhance control performance despite system delays.

This book brings together contributions from renowned authors in both academia and industry, reflecting a rich diversity of perspectives and expertise. As autonomous vehicles transition from research labs to public roads, the integration of perception and path planning systems becomes ever more critical. This book not only showcases technological breakthroughs but also inspires future research aimed at overcoming the remaining challenges. I am confident that this collection will serve as an invaluable resource for researchers, engineers, and decision-makers dedicated to advancing the frontier of autonomous vehicle technology.

Daniel Watzenig
Institute of Visual Computing
Graz University of Technology and CTO,
Virtual Vehicle Research
Graz, Austria

Contents

Scenario-Based Quantification of the Impact of Automated Vehicles on Traffic 1
Martin Sigl and Daniel Watzenig

Adversarial Examples in Environment Perception for Automated Driving 17
Jun Yan and Huilin Yin

Stable Resampling Strategies for Radar-Based Dynamic Occupancy Grids 69
Max Peter Ronecker, Michael Stolz, and Daniel Watzenig

Local Point Cloud Features for LiDAR Self-Supervised Representation Learning 89
Dušan Malić, Christian Fruhwirth-Reisinger, and Horst Possegger

On Torque-Vectoring Control for the Obstacle Avoidance Scenario: A Reinforcement Learning Case Study 107
Carmine Caponio, Mario Mihalkov, Umberto Montanaro, Patrick Gruber, Aldo Sorniotti, and Zoltan Hankovszki

Performance Evaluation of Path Tracking Controllers for Scaled Robotic Cars 127
Carmine Caponio, Pietro Stano, Giulio Scattolini, Nuccio Lo Bello, Michael Stickley, Mario Mihalkov, Alessandro Vigliani, Alessandro Rizzo, Aldo Sorniotti, and Umberto Montanaro

A Nonlinear Dead-Time Compensation Method for Path Tracking Control ... 155
Karin Festl and Michael Stolz

Scenario-Based Quantification of the Impact of Automated Vehicles on Traffic

Martin Sigl and Daniel Watzenig

Abstract

With the introduction of automated vehicles, it can be expected that new, previously unseen situations occur in hybrid traffic of both automated and human controlled vehicles. This impact of automated vehicles requires traffic metrics that quantify these changes. In this paper's state of the art section, such an impact quantification method is presented. It is then used as part of a new methodology that shows an impact analysis relying on splitting the operational design domain of automated vehicles into scenarios. A proof of concept shows that this split into sub-spaces is necessary for impact quantification of automated vehicles.

Keywords

Assessment of automated driving • Traffic quantification • Impact of automated vehicles

1 Introduction

Virtual assessment is the de facto state of the art for developing Automated Driving (AD) systems [6, 8]. Virtual assessment comes with advantages such as scalability, repeatability, and a safe and controllable test environment. Real-world tests remain an integral part of

M. Sigl (✉)
BMW AG, Munich, Germany
e-mail: martin.ms.sigl@bmw.de

D. Watzenig
Graz University of Technology, Institute of Visual Computing, Graz, Austria
e-mail: daniel.watzenig@tugraz.at

D. Watzenig (ed.), *Recent Advances in Autonomous Vehicle Technology–Perception and Path Planning*, Sustainable Mobility & Energy,
https://doi.org/10.1007/978-3-032-01001-8_1

the assessment process, focusing on validating and extending virtual tests. But the System Under Test (SUT) can only be efficiently investigated in virtual test environments, since any software change (or hardware change on the vehicle) invalidates past tests. Real-world test drives are therefore an important source for realistic test cases for virtual assessment.

Usually, virtual testing does not rely on precisely recreating an average drive to assess AD, for example like a daily commute of an employee driving into office. Instead, focus lies on a test-case oriented approach with somehow interesting content. In an AD assessment context, these test cases are also called scenarios [1]. Scenarios are meant to include all necessary information to assess the SUT. Next to the individual content requirements, scenarios shall cover the whole Operational Design Domain (ODD) in order to achieve reliable Automated Vehicle (AV) assessment. A logical scenario encapsulates the class or type of a test case, that usually would be described similarly, for example a cut-in or lane change. There is no limitation on the granularity or abstraction level of scenario descriptions, and human interpretability is not inherently necessary. In principle, a scenario is defined as a sequence of snapshots of the parts of a test case, which includes its static and dynamic objects and their behavior [12]. The goal of connecting test cases to scenarios and scenario classes is to structure the entirety of possible situations an AV can be in order to test that scenario space efficiently and effectively.

The foremost goal of AD assessment is guaranteeing the safety of the AD function. Since liability of driving maneuvers switches from driver to manufacturer from SAE automation level three an onwards, Original Equipment Manufacturers must be sure to validate their AD functions thoroughly before bringing them into market. Next to safety, the main goals of AD assessment include adherence to traffic rules and comfort. All these goals focus on the individual AV and its performance. An AV can achieve all these goals and still struggle with real-world applicability. AVs encounter other AVs and human drivers and must interact with each other. This changes existing interaction behavior which is normally seen between humans and also lead to previously unseen behavior and scenarios.

In Fig. 1, the additional situations introduced by AVs are depicted. The grey circle represents the human-only traffic. With the introduction of AVs, new scenarios will be introduced, depicted by a larger red ellipse that fully encapsulates the human-only traffic. The additional situations introduced may come from various sources. such as previously unseen AV maneuvers, or the mutual influence of human and AVs interaction.

Two examples of this will be explained in the following. The assumption of these examples is a highway domain with a certain fraction of AVs. First, AVs may tend to have a longer safety distance to preceding vehicles than human drivers do. This would decrease the number of vehicles on certain highway sections and decrease the overall traffic flow. This would be a negative effect on established human traffic behavior, ultimately leading to a decreased acceptance of AVs. In the second example, due to the main objective of safe operation, AVs may be hesitant to merge from highway ramps into traffic on highways. The most reliable safe state of the AV in that situation is to reduce velocity into standstill at the end of the merging lane and wait for a gap between oncoming vehicles that is deemed big enough to

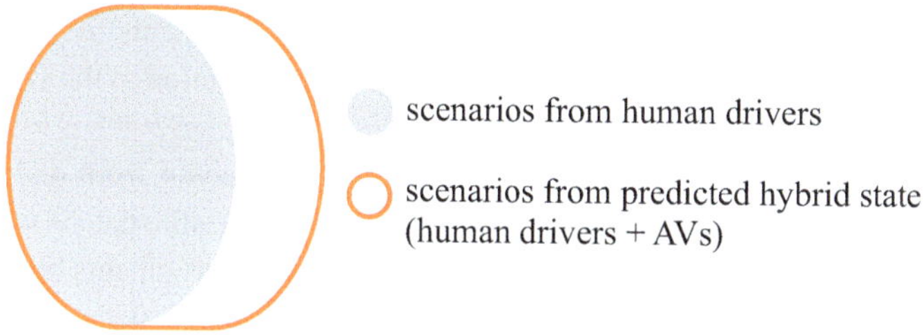

Fig. 1 The grey ellipses depicts all the possible scenarios and situations that human drivers can cause. Within a hybrid state with both human drivers and AVs, additional scenarios and situations can occur. This is depicted in the larger, red ellipses

safely merge onto the highway. This would lead to the previously unseen scenario, where AVs pile up on the end of merging lanes, causing subsequent effects and situations.

To investigate the change in existing traffic, metrics to quantify changes are necessary. For detecting new scenarios and investigating the hybrid traffic space, scenario space exploration methods are necessary. The combination of both tools will be the main focus of the upcoming sections, and ultimately describe scenario-based quantification of the impact of AV on traffic.

In Sect. 2, state of the art methods in AD and AV assessment are presented. These methods are then combined into a methodology for scenario-based impact analysis of AVs in Sect. 3. A proof of concept shows the applicability of the methodology with using lane changes as exemplary scenario in Sect. 4. The paper concludes with a discussion of the results (Sect. 5) and an outlook (Sect. 6).

2 State of the Art

2.1 Virtual Assessment of AD

Virtual assessment tools can be divided into two sub-types of simulating real-world test cases.

The first approach of relying on real-world test cases to derive scenarios is called *data-driven simulation* [7]. The validity of the scenario is given by a precise recreation of the real-world test. Key Performance Indicators (KPIs) that evaluate the AD function with regard to safety, adherence to traffic rules, and comfort are then applied on the virtualized version of the test case. This approach is expected to reduce overall effort to assess AD, since it yields sets of critical scenarios, which differs from real-world testing, where safety-critical scenarios are relatively rare [5, 16]. In an additional step, these virtualized versions of real-world tests can be varied. By changing behavior slightly, such as different velocity or movement profiles in a maneuver, the scenario space can be explored [1, 4]. A practical limit to this approach is the validity of the derived scenario. Since its source is a real-world test case, only this specific test case is valid. A change to its components by variation is only possible in a narrow range until it loses its grounding and connection to the valid real-world test case.

The second approach that relies on extracting realistic behavior from real-world traffic observations is called *statistical simulation*. Again, real-world traffic data is used as a basis. Behavioral information is extracted that describes how traffic participants behave in certain scenarios. This information is described with statistical methods such as distributions of velocities, accelerations, or distances between traffic objects. To generate an individual test case, samples from these distributions are drawn that parameterize the scenario for simulation (compare [15]).

While both approaches provide the necessary means to explore the scenarios space and therefore enable impact analysis of AVs, the statistical simulation has inherent ability to generate large, realistic data sets of traffic observations. It is therefore best suited to generate the foundation of scenario-based quantification of impact of AVs on traffic.

2.2 Traffic Quantification Metrics

Describing traffic quantitatively has been a basic tool for decades, for example in public road construction (compare [9]). In principle, those Traffic Quantification Metrics (TQMs) can be divided into two groups, *microscopic* and *macroscopic*. Microscopic TQMs describe the individual properties of traffic objects, such as velocities or other vehicle states. If they are aggregated over multiple traffic objects and time, they become macroscopic, such as traffic flow, for example. A typical, non-exhaustive set of TQMs can be seen in Table 1. How to investigate their ability to detect changes between traffic observations can be seen in [10]. In principle, any metric that somehow describes traffic quantitatively can be used for traffic impact analysis. The challenge lies in providing a method to evaluate TQMs that determines their sensitivity to changes in traffic. The following section presents such a method.

Table 1 Overview of TQMs used in the traffic space analysis

TQM	Description
v	Vehicle velocity
a	Vehicle acceleration
j	Vehicle jerk
THW	Time gaps between vehicles and their predecessor
TTC	TTC values between vehicles and their predecessor

2.3 Homogeneity and Heterogeneity Analysis

In principle, any TQM can be used for traffic impact analysis. Each TQM individually cannot deliver a holistic view on what is happening within a traffic observation. This section describes a method to evaluate and quantify traffic in its entirety by combining multiple TQMs into a single analysis approach based on [10]. For this analysis, the highD data set is used as a real-world traffic source [3]. The result of this approach is how similar or different traffic observations are. The evaluation criterion for TQMs therefore is the ability to quantify the *homogeneity* (similarity) or *heterogeneity* (difference) between traffic observations.

For this quantification process, Welch's t-test is used [14]. The choice of Welch's t-test over for example the Student's t-test (see [11]) is it's higher reliability of sample sizes and different variances of the samples. Both of the conditions can occur when comparing different traffic observations. The Welch's t-test's null hypothesis is the equality of the mean of the distribution of a TQM on a specific traffic observation:

$$H_0 : \mu(TQM_{TO_1}) = \mu(TQM_{TO_1}) \tag{1}$$

This comparison of the means is achieved by aggregating the TQM over all objects of a traffic observation. The challenge of testing lies in neither over- or underestimating (dis-)similarity. Having a stricter null hypothesis would lead to higher estimation of heterogeneity between two traffic observations, while a weaker hypothesis would overestimate homogeneity. The authors of [10] have shown that such a light-weight definition of similarity is balanced just enough to achieve a robust identification of homogeneity and heterogeneity between traffic observations.

Example: Comparison of Dissimilar Real-World Traffic Observations In Fig. 2, an exemplary heterogeneity analysis is depicted. It tests for the null hypothesis (with p-value at 0.03) by comparing two real-world traffic observations that are expected to differ significantly. In this case, two clusters have been derived from the highD highway data that is recorded with drones over german highways [3]. The properties of the two clusters are:

$$\begin{aligned} Cluster_1 :& \mu(TF) = 3338 \pm 178\ [veh/h] \\ & \mu(THW) = 0.65 \pm 0.25\ [1/s] \\ & \mu(v_{long.}) = 25.7 \pm 2.9\ [m/s] \\ Cluster_2 :& \mu(TF) = 1740 \pm 235\ [veh/h] \\ & \mu(THW) = 0.55 \pm 0.23\ [1/s] \\ & \mu(v_{long.}) = 25.75 \pm 3.6\ [m/s] \end{aligned} \tag{2}$$

The heterogeneity analysis assumes that the TQM evaluation shows dissimilarity for each TQM. Indeed, this is the case to all depicted TQMs. Besides this binary rejection or acceptance of the null hypothesis, it can be seen that some TQMs indicate a more uniform

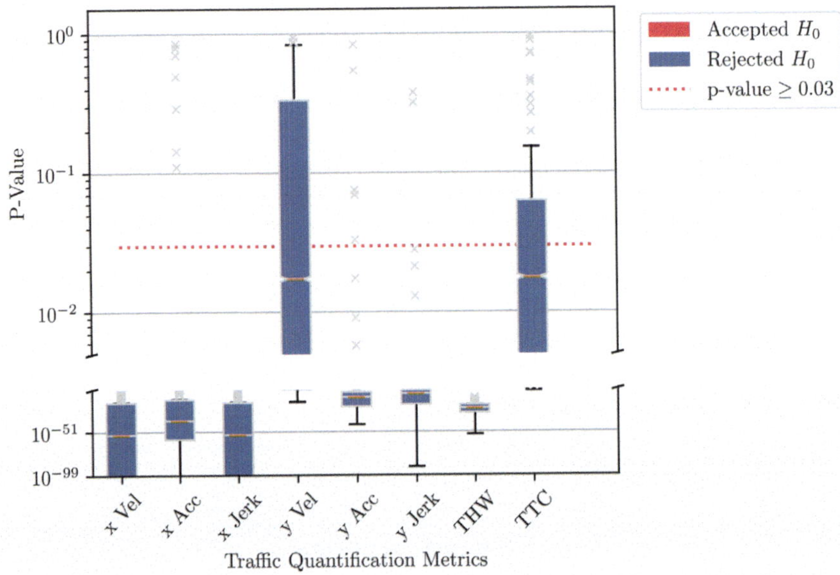

Fig. 2 Heterogeneity of traces between two real-world traffic observations. Longitudinal TQMs are depicted by the prefix 'x', lateral with 'y' respectively. Heterogeneity is assumed, and the blue pillars indicate that the null hypothesis is rejected, therefore the corresponding TQM indicates dissimilarity. Red pillars would show rejected heterogeneity, but to not occur in this analysis

behavior. This applies for the lateral velocity of objects and the TTC. This allows for the conclusion, that in traffic observations that otherwise differ very much, lateral movement and the distance to preceding vehicles is more similarly distributed than other TQMs. An explanation would be that lateral movement and safety distances are more independent from general traffic conditions as other traffic object behavior.

Example: Validation of Simulation with Homogeneity Analysis For the simulated traffic observations, the simulation framework openPASS is used, which generates traffic observations based on the Monte-Carlo method [2]. In Fig. 3, an exemplary homogeneity analysis is depicted. It tests for the null hypothesis (with p-value at 0.03) by comparing a simulated traffic observation with a real-world traffic observation. In the examined case, a statistical simulation tries to recreate real-world traffic using the above mentioned process to derive behavior of traffic objects and sampling virtual traffic objects. This also shows one of the abilities of the homogeneity analysis approach, where the expected homogeneity property of all TQMs are an indicator of the simulation's ability to precisely recreate the real-world traffic.

In the depicted example, the null hypothesis is accepted for many of the TQMs, except for longitudinal and lateral acceleration values of the traffic objects. In this case, the analysis shows deficiencies in either the derivation of acceleration behavior of traffic objects from

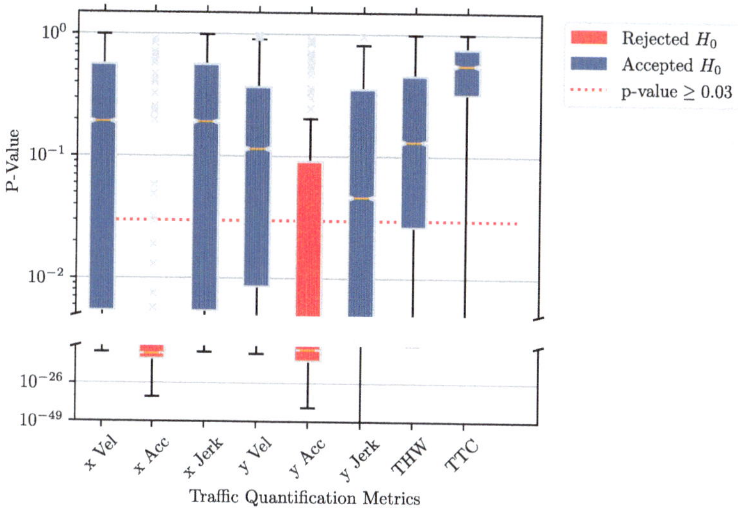

Fig. 3 **Homogeneity** of traces between a real-world and a simulated traffic observation. Longitudinal TQMs are depicted by the prefix 'x', lateral with 'y' respectively

the real world, or in recreating the behavior within the simulation. This information can be used to targeted improvements of the statistical simulation chain. The next section will show how the homogeneity analysis can contribute to impact analysis of AVs.

2.4 Impact Analysis of AVs

In principle, all necessary tools of impact analysis of AVs are available. Either data-driven or statistical simulation deliver virtual traffic observations that can include single or multiple AVs. The homogeneity/heterogeneity analysis then delivers insight on which traffic object behavior changes between different traffic observations. The presented means allow to compare three basic types of traffic observations:

1. real-world traffic with human only drivers
2. simulated traffic recreating real-world traffic
3. simulated traffic with AVs.

The first traffic observation type with only human drivers serves as the baseline or ground truth for the following types. Transforming it into the virtual world is the prerequisite for virtual assessment of AD. By comparing these two traffic observation sources, the homogeneity analysis provides means to validate the transformation ability of simulation processes. After incorporating a fraction of AVs into the simulation, traffic observations with and without

AVs can be compared. With the homogeneity analysis, differences are detected. This gives a first insight on which TQMs can detect changes introduced by AVs. The actual quantification is then achieved by looking at the values that define the specific TQM that has changed. This can reach from looking at mean and standard variance of the respective TQM, but also on median values or outliers. For example, a mean longitudinal velocity that falls with simultaneously decreasing variance, this indicates slower, but smoother traffic flow. Nevertheless, several questions and problems are still open, which will be discussed in the following section.

2.5 Interpreting Traffic Clusters

Looking at whole traffic observations for the heterogeneity and homogeneity analysis comes with a disadvantage. Changes in traffic object behavior may only occur in rare situations and can easily be lost in vast data analysis processes. In the previously mentioned scenario of merging onto a highway leading to many AVs stopping at the end of the merging lane may be significant for that type of scenario. But this scenario only makes up a small fraction of overall behavior of traffic on highways. An example from [10] takes lane changes as such an exemplary traffic situation and will be presented in Sect. 4. To reliably detect those changes, the investigated traffic observations must be divided into sub-spaces that are more sensitive to different traffic object behavior. In principle, scenarios fulfill the requirements to divide traffic observations into sub-spaces. Although their approach usually includes having a logically derived situation class rather than splitting up traffic observations into a union of more abstract scenarios.

In contrast to (logical) scenario-based clustering, more abstract clustering methods might rely on similarity metrics that cannot easily be interpreted by humans. In combination with the presented homogeneity analysis with TQMs, this interpretability can be reached independently of the underlying clustering method. Proper clustering is necessary to proof the coverage of assessment of AD and AVs, and methods to do so already exist. One of them is presented in the next section.

2.6 Exploration of Scenario Space

In [4], a method for the exploration of the scenarios space to achieve full coverage of AD assessment is presented. It consists of three stages (compare with Fig. 4):

1. **Discoverage** of new scenario clusters by similarity of scenarios
2. **Expansion** of cluster with variations of similar scenarios
3. **Convergence** of the clusters, when no new clusters are discovered.

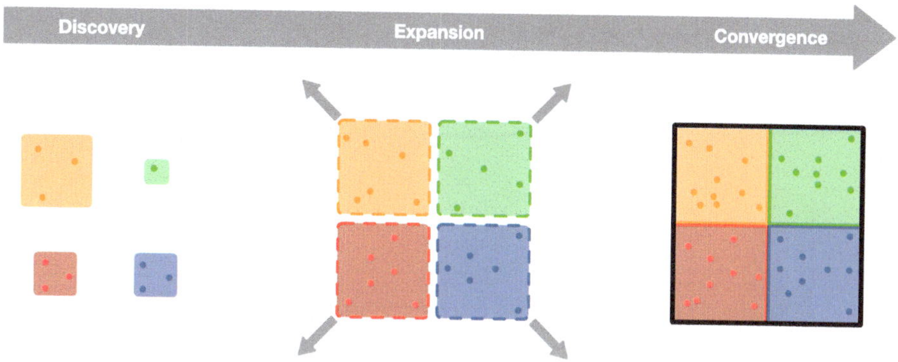

Fig. 4 Schematic depiction of the three steps of scenario space exploration. The squares depict scenario clusters, while the little dots within represent single scenarios. Expansions increases cluster sizes by assigning new scenarios to clusters, finally achieving convergence

In this approach, a similarity metric is used to classify scenarios of traffic observations. The spatio-temporal development of the objects is the main contributor to this similarity metric, resulting in a distance between two scenarios. This distance is used to define the membership of a scenario to a certain cluster. This discovery step is performed until there are no scenarios left in the traffic observation source. Expanding these cluster afterwards is the next step. Since the original traffic observation source is likely to not include all possible variants of scenarios, these variants have to be generated. This local exploration of scenarios must focus on keeping the validity of the variated scenario, but is otherwise free regarding the used method [13]. The expansion step will at some point saturate, where the clusters converge and no new clusters are discovered. Reaching this goal indicates that the complete scenario space is contained in this set of scenarios.

3 Improvements in Impact Analysis

Scenario-based quantification of the impact of AVs on traffic requires the combination of the presented state-of-the-art methods. Figure 5 shows the basic methodology. The source of the methodology is a representation of hybrid traffic with AVs. This source is preferably coming from simulation, due to it's advantages described earlier. However, this is not a hard requirement, the data source can come from real-world traffic observations as well. For proper evaluation, a baseline traffic observation with human-only traffic is necessary. Again, this data can come from either a virtual or real-world source. These data sources correspond to the red and grey ellipses of Fig. 1.

The hybrid traffic observation with AVs then is clustered and explored. The exploration step is necessary, since new clusters of scenarios, or at least a different composition of

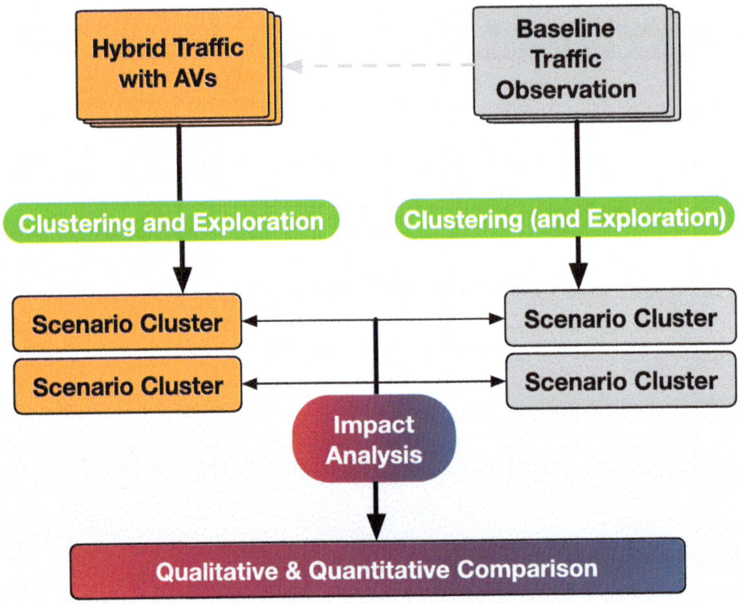

Fig. 5 Methodology for scenario-based quantification of impact of AVs on traffic

clusters can be expected to be introduced by the new behavioral patterns of AD functions. The exploration step is best performed in virtual testing environments, since they enable variation of existing scenarios much more easily as real-world data sources. Clustering is also performed for the baseline traffic observation. It is imperative to cluster traffic observations before deriving any statements on what behavior they contain, since changes from one cluster can nullify changes of another cluster. Also, the differing relative occurrences of certain events also inhibit reliable interpretation of changes when they are hidden by a vast majority of relatively unchanged traffic situations. Exploration is also necessary, if the baseline is not containing the complete traffic space, including all of the possible scenarios.

Another requirement is that the clustering mechanism is performed the exact same way for both the hybrid traffic with AVs as well as the baseline traffic observation. Although depicted as two separate steps in the methodology figure, this step must ensure that the same clustering criteria are applied independently of the scenario source. The final impact analysis is then performed between similar clusters and the cluster structure itself.

The overall structure of clusters delivers the first quantitative analysis result. Different sizes of clusters are induced by higher or lower frequencies of that scenario happening. For example, the scenario cluster of lane changes might be containing a lot more scenario instances in one of the two scenario sources. This would then indicate that AVs cause a shift in commonness of that scenario compared to the human-only traffic space.

If there is no corresponding cluster originating from the simulated AV traffic space present in the baseline traffic observation, this stand-alone cluster contains traffic situations newly introduced by AVs. These clusters are particularly important for AD assessment, since their non-existence in baseline traffic observations leads to these scenarios being overlooked in assessment strategies. Practical real-world examples are deadlocks caused by multiple AVs interacting with each other. This is a scenario that cannot be derived as possible test case from real-world, human-only traffic observations. The discovery of such previously unseen scenarios is indispensable before large-scale market introduction of AVs.

The homogeneity and heterogeneity analysis delivers qualitative statements about clusters. The TQMs of the method show discrepancies of behavior between clusters. Additionally, this method reveals shortcomings or blind spots of clustering mechanisms, since it offers insight regarding inner-cluster similarity and inter-cluster differences.

The final step of quantifying the impact of AVs consists of applying validated TQMs on corresponding clusters from each scenario source. Looking at the distribution of TQMs reveals the exact differences between those two clusters. This might include the requirement for scenario-based TQMs as well. In the exemplary case of a lane-change cluster, a TQM must be able to detect changes in the behavior that is under investigation. The introduction of new TQMs can be again evaluated with the homogeneity analysis, revealing a TQM's ability to detect changes between traffic observations. Ultimately, the presented method leads to precise quantification of AV impact based on scenario clusters.

In the following section, this methodology is applied on two different traffic observations. Both observations come from the real-world data source HighD, to offer the applicability of the method independent from distortions and errors that could be introduced by simulations frameworks. Also, the HighD data does not contain AVs, therefore showing human-only traffic. This constraint is mitigated by using the two dissimilar traffic clusters of HighD that were described earlier.

4 Proof of Concept: Lane Change

As an exemplary application of the previously presented methodology, an impact analysis of different sets of lane change observations is depicted in this section. The analysis will be twofold. First a homogeneity analysis between the whole traffic space and the scenario cluster will be presented. Second, a homogeneity analysis between to scenario clusters of different parent traffic observations.

Lane changes serve as the investigated scenario cluster, because they are relatively easy to detect by changes of lane IDs in the data and can be interpreted straightforward by humans.

4.1 Parent Observation Versus Scenario Cluster

For this analysis, a scenario cluster is derived from the HighD observation $Cluster_1$. Figure 6 shows the results of the homogeneity analysis between parent observation and scenario

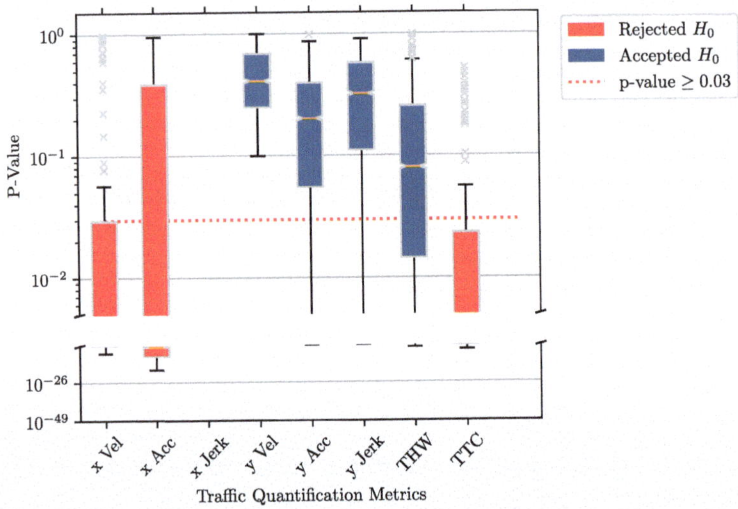

Fig. 6 This homogeneity test compares the whole traffic observation of $Cluster_1$ with its lane change scenario cluster. Some of the TQMs fail the similarity test, since lane changes do not share all the same properties as the whole traffic observation

cluster. Some of the TQMs indicate dissimilarity between the two sets of traffic data while other indicate similarity. This shows how some properties are inherited from the whole parent observation, while others differ. Proper clustering expects that the individual cluster is somehow cutting out data with peculiarities that are not present in all parts of the original data set.

Applying this homogeneity analysis to different clusters of the parent data set can offer insight to the quality of the used clustering technique. However, this cannot be use as verification of the actual clustering, since individual clusters can happen to show similar TQMs by chance and TQMs are not necessarily a set of metrics that capture all properties of a traffic observation.

In conclusion, this homogeneity analysis provides insight to the similarity of the scenario cluster and its parent traffic observation. In the case of the lane change scenario cluster, the expectation of some level of dissimilarity is reached. For the next step, scenario clusters derived from different traffic observations are investigated regarding their homogeneity.

4.2 Scenario Clusters of Different Traffic Observations

For this analysis, a second lane change scenario cluster is derived from HighD $Cluster_2$. It is then compared in a homogeneity analysis with the lane change scenario cluster from HighD $Cluster_1$. The results are depicted in Fig. 7. The similarity between the two scenario

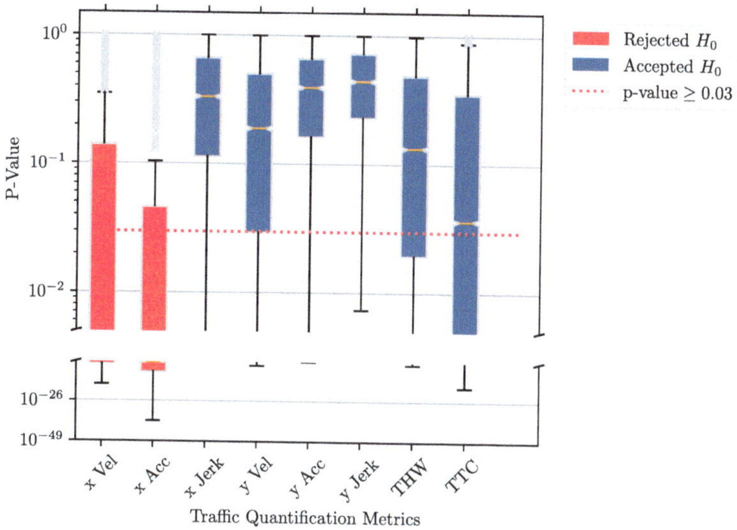

Fig. 7 Homogeneity test between lane change scenario clusters of $Cluster_1$ and $Cluster_2$. The test passes regarding most TQMs, because the scenario clusters are relatively similar, even across dissimilar traffic observation clusters. The actual AV impact analysis happens in the bottom row

clusters is significantly higher than between the two whole traffic observations (compare Fig. 2). In the heterogeneity analysis of Fig. 2, all TQMs reject the Null hypothesis of similarity. The scenario clusters however show similarities for most of the TQMs except longitudinal velocity and acceleration.

In conclusion, this type of homogeneity analysis shows what traffic object behavior (or TQM) is similar under different circumstances (or traffic observations). The results of this section play a crucial role in scenario-based impact analysis, which will be explained further in the next section.

5 Discussion of Scenario-Based Impact Analysis of AVs

Revisiting the methodology of Sect. 3, this part takes a closer look at the impact analysis of Fig. 5. The specific results of the proof of concept in Sect. 4 are transferred to the overall goal of scenario-based impact analysis of AVs.

Figure 8 shows a schematic depiction of different possible homogeneity analyses. Instead of two traffic observations from human-only traffic as in the proof of concept section, it assumes one source to come from hybrid traffic including AVs. In the first row of the schematic, a homogeneity analysis between the hybrid and baseline traffic observation is depicted. As previously stated, it can give some insight, but it can be assumed that most of the interesting behavior in sub-spaces of the traffic observation is masked by the sheer

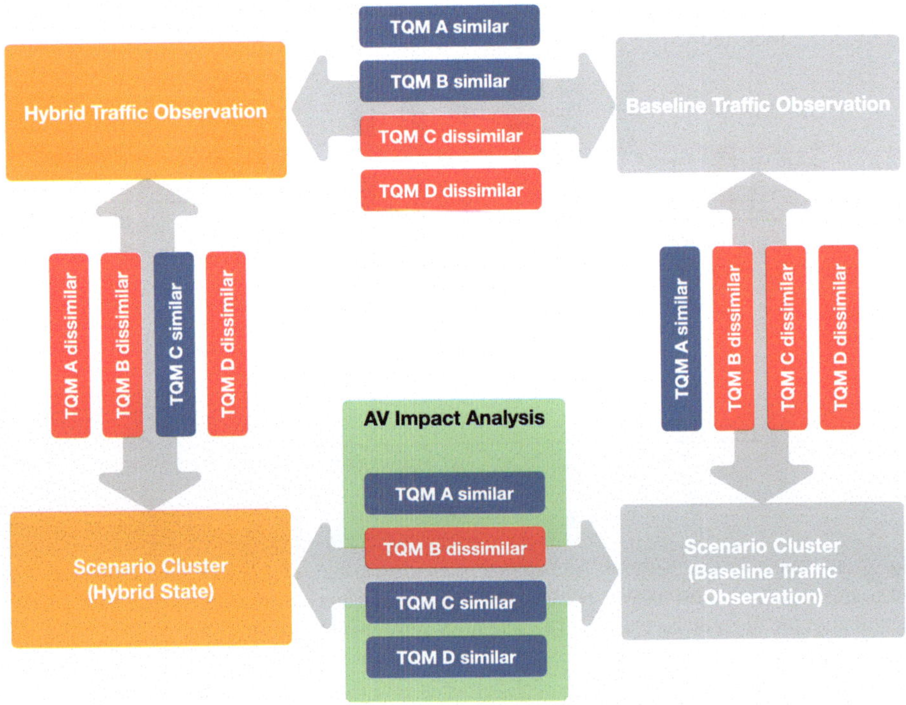

Fig. 8 Schematic depiction of different homogeneity analyses. Regarding an individual scenario cluster and its parent traffic observations, four homogeneity analyses can be performed

amount of other situations within the data sets. TQMs can show similarity or dissimilarity, but the effects of AVs on that change can not be attributed to specific changes in certain situations directly.

The bottom row shows the specific scenario clusters of each traffic observation. Similar to the proof of concept section, an homogeneity analysis between a scenario cluster and its parent traffic observation can be performed to gain insight on the clustering method. Again, this does not directly allow a statement on the impact of AVs.

The most important part of the analysis is in the comparison of the two scenario clusters in the bottom row. As shown in the proof of concept, scenario clusters tend to show predominantly similarities. This already was the case with the otherwise dissimilar parent traffic observations used in the proof of concept. A comparison between more similar baseline and hybrid traffic is expected to show more similarities in the parent traffic observations. The expected result of analyzing the scenario clusters is an even more homogeneous set of TQMs. However, the focus lies on the dissimilar TQMs of comparing scenario clusters from the different sources. These dissimilarities show significant changes that are introduced by hybrid traffic, and therefore AVs, compared to the human-only baseline. This qualitative statement of AVs directly influencing the structure and behavior of traffic can also be quantified by looking at the actual values of the TQMs. These quantitative metrics consist of a TQM's mean and variance, but can also include more sophisticated types of analysis of the underlying distribution of a TQM in the specific scenario cluster.

6 Future Work

While the basic methodology to perform scenario-based quantification of the impact of AVs on traffic is explained in the previous sections, several questions and challenges must be answered in future work. By dividing whole traffic observations into scenario clusters, scenario-based TQMs become necessary. For logical scenarios such as lane changes or similar, TQMs can be derived from expert knowledge and from expected behavioral metrics. For abstract clustering mechanisms this is significantly more complicated and needs further investigation.

Also, this work omits the frequency of scenarios. The scenario cluster sizes can vary between hybrid and baseline traffic observations, indicating a shift of how often situations occur because of AVs. Future work has to tackle these challenges to further improve the ability to analyze the impact of AVs in hybrid traffic.

References

1. Groh, K., Kuehbeck, T., Fleischmann, B., Schiementz, M., Chibelushi, C.C.: Towards a Scenario-Based Assessment Method for Highly Automated Driving Functions. (2017)
2. Guindon, C.: OpenPASS Working Group. The Eclipse Foundation (2017)
3. Krajewski, R., Bock, J., Kloeker, L., Eckstein, L.: The highD Dataset: A Drone Dataset of Naturalistic Vehicle Trajectories on German Highways for Validation of Highly Automated Driving Systems. In: 2018 21st Int. Conf. Intell. Transp. Syst. (ITSC), pp. 2118–2125. IEEE (2018)
4. Kerber, J., Wagner, S., Groh, K., Notz, D., Kühbeck, T., Watzenig, D., Knoll, A.: Clustering of the Scenario Space for the Assessment of Automated Driving. In: 2020 IEEE Intelligent Vehicles Symposium (IV), pp. 578–583 (2020)
5. Lawitzky, A., Althoff, D., Passenberg, C.F., Tanzmeister, G., Wollherr, D., Buss, M.: Interactive scene prediction for automotive applications. In: Intelligent Vehicles Symposium (IV), 2013 IEEE, pp. 1028–1033. IEEE (2013)
6. Maurer, M., Gerdes, J.C., Lenz, B., Winner, H. (eds.): Autonomes Fahren: Technische, rechtliche und gesellschaftliche Aspekte. Springer Vieweg (2015)
7. Minnerup, P., Kessler, T., Knoll, A.: Collecting Simulation Scenarios by Analyzing Physical Test Drives. In: 2015 IEEE 18th Int. Conf. Intell. Transp. Syst., pp. 2915–2920 (2015)
8. Riedmaier, S., Ponn, T., Ludwig, D., Schick, B., Diermeyer, F.: Survey on scenario-based safety assessment of automated vehicles. IEEE Access **8**, 87456–87477 (2020)
9. Smeed, R.J., Bennett, G.T.: Research on road safety and traffic flow. In: Road Engineering Division (1963)

10. Sigl, M., Prins, B., Schütz, C., Wagner, S., Schulte, F., Watzenig, D.: Quantifying the impact of automated vehicles on traffic. In: 2023 IEEE International Automated Vehicle Validation Conference (IAVVC). IEEE (2023)
11. Student: The probable error of a mean. Biometrika, 1–25 (1908)
12. Ulbrich, S., Menzel, T., Reschka, A., Schuldt, F., Maurer, M.: Defining and Substantiating the Terms Scene, Situation, and Scenario for Automated Driving. In: 2015 IEEE 18th Int. Conf. Intell. Transp. Syst., pp. 982–988 (2015)
13. Wagner, S.K.: Efficient scenario-based assessment of automated driving systems through virtual testing. (2021)
14. Welch, B.L.: The generalization of 'student's' problem when several different population variances are involved. Biometrika **34**(1–2), 28–35 (1947)
15. Wang, L., Fahrenkrog, F., Vogt, T., Jung, O., Kates, R.: Prospective safety assessment of highly automated driving functions using stochastic traffic simulation. (2017)
16. Wagner, S., Groh, K., Kühbeck, T., Knoll, A.: Towards Cross-Verification and Use of Simulation in the Assessment of Automated Driving. In: 2019 IEEE Intelligent Vehicles Symposium (IV), pp. 1589–1596 (2019)

2 Adversarial Examples in Environment Perception for Automated Driving

Jun Yan and Huilin Yin

Abstract

The renaissance of deep learning has led to the massive development of automated driving. However, deep neural networks are vulnerable to adversarial examples. The perturbations of adversarial examples are imperceptible to human eyes but can lead to the false predictions of neural networks. It poses a huge risk to artificial intelligence (AI) applications for automated driving. This survey systematically reviews the development of adversarial robustness research over the past decade, including the attack and defense methods and their applications in automated driving. The growth of automated driving pushes forward the realization of trustworthy AI applications. This review lists significant references in the research history of adversarial examples.

Keywords

Deep learning • Machine vision • Automated driving • Adversarial examples • Robustness

---

This review is a section of the "Automated Driving Vehicle Technologies" book.

J. Yan · H. Yin ((✉))
School of Electronic and Information Engineering, Tongji University, Shanghai, China
E-Mail: toni.susin@upc.edu

J. Yan
E-Mail: yanjun@tongji.edu.cn

D. Watzenig (ed.), *Recent Advances in Autonomous Vehicle Technology–Perception and Path Planning*, Sustainable Mobility & Energy,
https://doi.org/10.1007/978-3-032-01001-8_2

1 Introduction

Deep learning has been hugely successful over the past decade, powered by Graph Processing Units (GPUs), big data, and human intelligence. This success has spurred an artificial intelligence (AI) renaissance, enabling amazing applications like chat assistants, embodied robotics, and autonomous driving. In the past decade, AI technologies have matured enough for autonomous driving to be productized. The Society of Automotive Engineers (SAE) categorizes driving levels from no automation (Level 0) to full automation (Level 5), aiming for fully autonomous driving [1]. With AI advancements, some countries have introduced decrees for testing fully autonomous systems.

However, deep neural networks are vulnerable to adversarial attacks [3–6] where slight raw data perturbations fool networks into wrong predictions. Many pieces of research have delved into the exploration to promote AI security, including attacks [3–5], defenses [6, 7], systematic evaluations [8–10], and interpretations [11, 12]. Figure 1 shows huge research interest in adversarial robustness, although valuable scientific problems may be saturated. This saturation validates the impact and significant of the adversarial robustness research.

The issue of adversarial examples relates to the security risks of cyber-physical systems. In these systems, adversarial examples can serve as malicious data in the Internet of Vehicles. For instance, an automated vehicle's decision-making system can be misled by maliciously modifying a traffic sign. These adversarial examples reveal AI systems' vulnerabilities, inducing wrong decisions and potential security risks. Understanding adversarial examples' characterization and creation is critical for designing secure cyber-physical systems and ensuring the information systems' robustness. Meanwhile, other security and privacy risks also harm AI technologies' trustworthiness. Figure 2 shows a macro story of AI technology

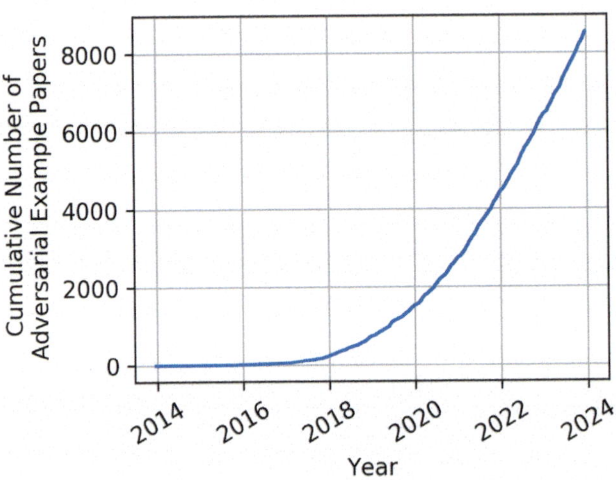

Fig. 1 The growth of papers related to adversarial robustness [2] @Nicholas Carlini's Blog

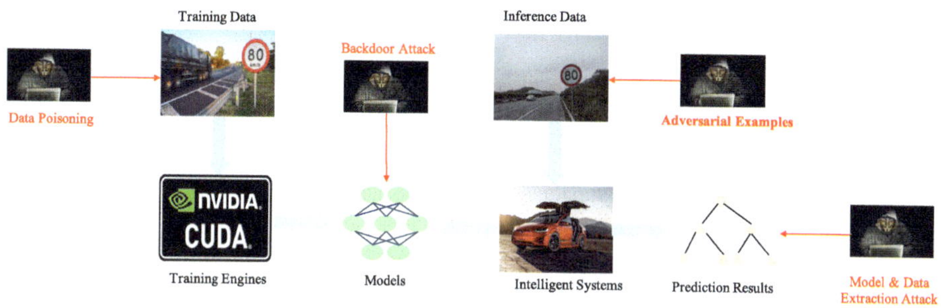

Fig. 2 Deep learning systems and the encountered attacks. Adversarial attacks happen in the model prediction process

risks. Adversarial examples primarily threaten during the inference phase of AI systems. Despite the importance of paying attention to other AI risks, this survey mainly reviews the research history of adversarial examples.

Giving attention to adversarial robustness research is critical to advancing the development of intelligent vehicles. The existence of adversarial examples poses a huge threat to the tasks of automated driving, including traffic sign recognition [13], vehicle detection [14], trajectory prediction [15], LiDAR perception [16], lane segmentation [17], and SLAM (Simultaneous Localization and Mapping) [18]. Figure 3 demonstrates the adversar-

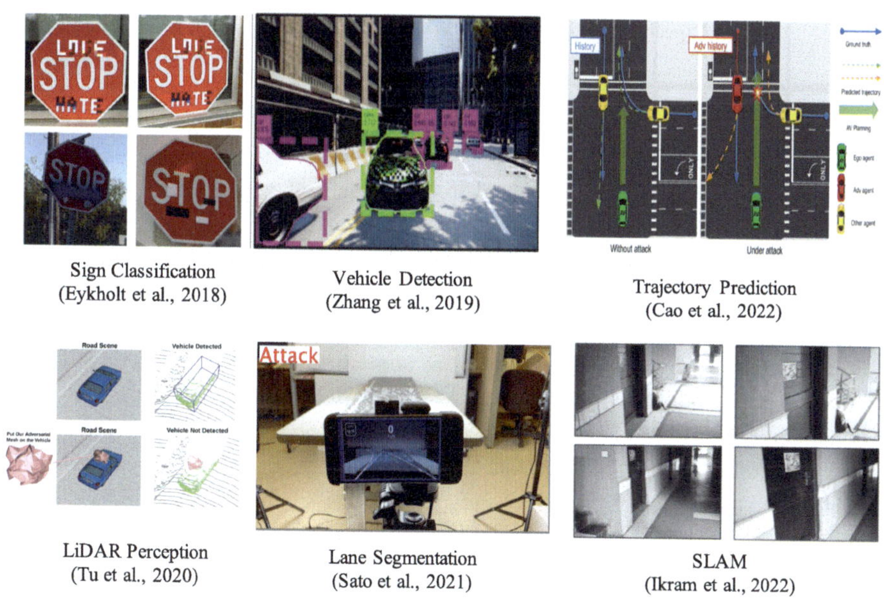

Fig. 3 The threat of adversarial examples in the practical tasks of automated driving

Table 1 Notations and explanations

Symbol	Description
x_0	A clean sample
$\tilde{x_i}$	An adversarial example
$\mathbf{x}$	The vectors of clean data
$\mathbf{y}$	The vectors of labels
$\tilde{\mathbf{x}}$	The vectors of adversarial examples
ϵ	The perturbation budget
$\mathcal{B}(x_0, \epsilon)$	The neighborhood ball of x_0 with radius of ϵ
x'	A sample in $\mathcal{B}(x_0, \epsilon)$
$\mathbf{x}'$	The vector sample in $\mathcal{B}(x_0, \epsilon)$
y_0	The groundtruth label
$\mathbf{I}$	The unit matrix
$\nabla_z J(\cdot)$	The gradient of a scalar function J with respect to $\mathbf{z}$
$f_{\mathbf{\Theta}}$	The neural network function with the parameter space $\mathbf{\Theta}$
δ	The added perturbation
$\|\boldsymbol{x}\|_p (p \geq 1)$	The vector p-norm of $\boldsymbol{x} = [x_1 \|, \ldots, x_d]$, defined as $\|x\|_p = \left(\sum_{i=1}^{d} \|x_i\|^p\right)^{1/p}$
$\|x\|_\infty$	infinity norm of $\boldsymbol{x} = [x_1 \|, \ldots, x_d]$, defined as $\|x\|_\infty = \max_{i \in [d]} \|x_i\|$
$\mathcal{L}(\cdot)$	The adversarial loss function
K	The number of classes in a classification task
L	The Lipschitz constant
$\odot$	Hadamard product

ial vulnerability risk towards the automated driving, which indicates the necessity to advance the research on adversarial robustness.

This review introduces adversarial examples' theories, methods, and automated driving applications. In particular, it focuses on the adversarial examples related to the environment perception systems for automated driving. Section 2 interprets fundamental concepts. Section 3 describes representative adversarial attack methods. Section 4 recalls adversarial defense methods. Section 5 introduces adversarial examples in automated driving applications. Section 6 highlights adversarial examples' relationship to the Safety of the Intended Functionality (SOTIF). Section 7 provides a future outlook for adversarial robustness research. Finally, Sect. 8 concludes the review.

2 Theory Preliminary

This section reviews the essential concepts and theory preliminary in the adversarial examples research. Table 1 gives essential notations which would be utilized to illustrate the studies. The following subsections review the mechanisms of gradient-based adversarial attacks, adversarial training, and randomized smoothing.

2.1 Gradient-Based Adversarial Attacks

Given the original examples $\mathbf{x_0}$, a successful adversary aims to find the relative adversarial examples $\tilde{\mathbf{x}} = \mathbf{x_0} + \boldsymbol{\Delta}$ that can deceive the visual system with the small perturbation $\boldsymbol{\Delta}$. It is a constrained optimization problem that the adversarial examples locate in the norm sphere of original examples defined in Eq. (1):

$$\underset{\tilde{\mathbf{x}}}{\operatorname{argmax}} f(\tilde{\mathbf{x}}) \text{ s.t. } \tilde{\mathbf{x}} \in \mathcal{B}(\mathbf{x}_0), \tag{1}$$

where $f(\tilde{\mathbf{x}})$ denotes the deceit on the classifier function $f(.)$ and $\mathcal{B}(\mathbf{x}_0)$ is a small region (norm sphere) with the adversarial perturbations. The norm sphere $\mathcal{B}(\mathbf{x}_0)$ is also a constrained set that the ℓ_p norm can measure the distance defined in Eq. (2):

$$\mathcal{B}(\mathbf{x}_0) = \left\{\tilde{\mathbf{x}} : \|\tilde{\mathbf{x}} - \mathbf{x}_0\|_p \leq \epsilon\right\}. \tag{2}$$

An original groundtruth label is y_0. There exist two types of attack categories. If the adversarial example $\tilde{x}_i$ belongs to a specific class y_t, this is a targeted attack defined in Eq. (3):

$$\underset{\tilde{x}_i}{\operatorname{argmax}} f(\tilde{x}_i) = y_t. \tag{3}$$

This means that the adversary induces the model to misclassify the data as the specific wrong label. Otherwise, it is an untargeted attack defined in Eq. (4):

$$\underset{\tilde{x}_i}{\operatorname{argmax}} f(\tilde{x}_i) \neq y_0. \tag{4}$$

In such a scenario, the adversary induces the model to misclassify the data as the unspecific wrong label.

The gradient-based attack is a ℓ_∞-norm steepest descent attack. The adversary utilizes a linear approximation of the objective function to search for the perturbations. Assumed that $g(.)$ is an attack procedure, $\mathbf{x}$ and $\mathbf{x}'$ is a batch of clean data and the searched examples in the neighborhood of $\mathbf{x}$. Eq. (5) defines the linear approximation of first-order Taylor expansion.

$$g(\mathbf{x}') \approx g(\mathbf{x}) + \nabla g(\mathbf{x})^T \left(\mathbf{x}' - \mathbf{x}\right). \tag{5}$$

A closed-form solution defined in Eq. (6) of this constrained optimization problem can be derived:

$$\mathbf{x}' = \mathbf{x} + \epsilon \operatorname{sign}(\nabla g(\mathbf{x})). \tag{6}$$

The symbol sign$(\cdot) \in \{+1, -1\}$ denotes element-wise sign values. Equation (6) is a mathematical form of the fast gradient sign method (FGSM), which is a milestone work of AI security. The FGSM attack pipeline can realize both untargeted and targeted attacks. If the adversary runs the FGSM method for multiple iterations T, Eq. (7) can be deduced:

$$\boldsymbol{x}_{t+1} = \Pi(\boldsymbol{x}_t + \alpha \operatorname{sign}(\nabla g(\boldsymbol{x}_t))) \quad \forall t = 0, \ldots, T-1, \tag{7}$$

where α is a step size for gradient-based attack and $\mathbf{x}_t$ is the perturbed data in the time step t. The multiple-step attack would terminate if the sampled data of adversarial examples $\tilde{\mathbf{x}}$ is found. Equation (7) defines the mathematical equation of iterative FGSM (I-FGSM) attack [19] and projected gradient descent (PGD) attack [6]. Compared to the I-FGSM attack, the PGD attack projects the perturbed input back onto the set of allowable inputs to ensure the modified image still has pixel values in the valid range. Currently, the PGD attack is a de facto effective gradient-based attack in the practical application.

2.2 Adversarial Defense

The Section of Theory Preliminary would mainly focus on two effective defense methods under the adversarial attacks: adversarial training [6] and certified randomized smoothing [20]. Other methods will be introduced in the next paragraphs.

The adversarial training procedure aims to minimize the expected empirical risk while maximizing the adversarial perturbations. Equation (8) defines the form of adversarial training:

$$\arg\min_{\theta} E_{(\mathbf{x},\mathbf{y})\in\mathcal{D}} \max_{\mathbf{x}'\in\|\mathbf{x}'-\mathbf{x}\|_\infty\le\epsilon} L\left(f\left(\mathbf{x}'\right), \gamma\right), \tag{8}$$

where E denotes expectation, $(\mathbf{x}, \mathbf{y}) \in \mathcal{D}$ denotes the data samples and their labels randomly drawn from the distribution $\mathcal{D}$, $L(.)$ is a supervised loss function, f is the fitting function of neural networks, and $\mathbf{x}'$ is the symbol of the perturbed data of ϵ-ball. Adversarial training is an effective shield against the adversarial attacks to handle the crisis of model leakages or transfer-based attacks.

Beside the empirical defense such as adversarial training, the defenders also desire to compute the "certified radius" [20], where it provides a robustness guarantee with a high probability that any perturbation within such radius will give a robust prediction. Figure 4 describes the mechanism of certified randomized smoothing. Assumed that $g(.)$ is a smoothed classifier with the base classifier $f(.)$ under the adversarial attacks, where the data is perturbed by isotropic Gaussian noise, there exists a formulation defined in Eq. (9):

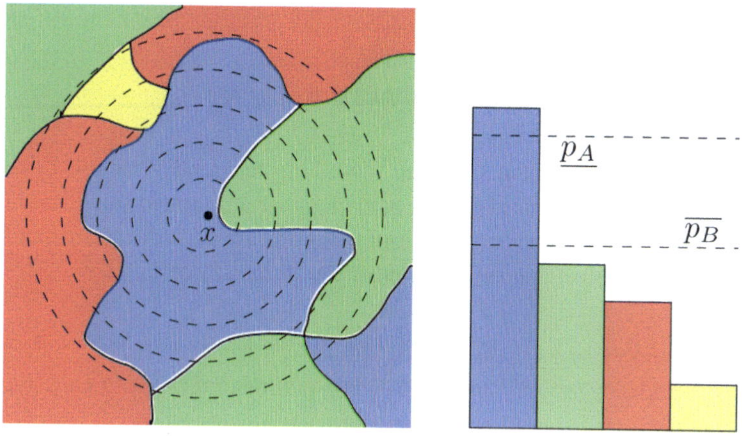

Fig. 4 The evaluation paradigm of smoothed classifier [20]. **Left**: the decision regions of the base classifier f are marked in different colors. The dotted lines represent the level sets of the distribution $\mathcal{N}\left(x, \sigma^2 I\right)$. **Right**: the distribution $\mathcal{N}\left(x, \sigma^2 I\right)$. Here, $\underline{p_A}$ is a lower bound on the probability of the top class, and $\overline{p_B}$ is an upper bound on the probability of each other class. The classifier function is in blue

$$
\begin{aligned}
g(x) &= \arg\max_{c \in \mathcal{Y}} \mathbb{P}(f(x+\varepsilon)=c) \\
\text{where } \varepsilon &\sim \mathcal{N}\left(0, \sigma^2 I\right)
\end{aligned}
. \tag{9}
$$

Assume that when the base classifier $f(.)$ categorizes a sample drawn from the distribution $\mathcal{N}\left(x, \sigma^2 I\right)$, it returns the most probable class c_A with the probability p_A. Simultaneously, the second most probable class, referred to as the "runner-up," is returned with a probability of p_B. The smoothed classifier is robust within the ℓ_2 radius $R = \frac{\sigma}{2}\left(\Phi^{-1}\left(p_A\right) - \Phi^{-1}\left(p_B\right)\right)$, where Φ^{-1} is the inverse of the standard Gaussian CDF (Cumulative Distribution Function). This review will highlight two important theorems of certified randomized smoothing.

Theorem 1 is a crucial result. The certified robustness can be built on the neural network models, if any, are satisfied by modern deep architectures. The certified radius R tends to be large under the following conditions: (1) the noise level σ is high; (2) the probability associated with the top class c_A is high; and (3) the probabilities corresponding to all other classes are low.

Theorem 1 *[20] Let $f : \mathbb{R}^d \to \mathcal{Y}$ be be any deterministic or random function, and let $\varepsilon \sim \mathcal{N}\left(0, \sigma^2 I\right)$. Let g be defined as in Eq. (9). Suppose $c_A \in \mathcal{Y}$ and $\underline{p_A}, \overline{p_B} \in [0, 1]$ satisfy:*

$$
\mathbb{P}\left(f(x+\varepsilon) = c_A\right) \geq \underline{p_A} \geq \overline{p_B} \geq \max_{c \neq c_A} \mathbb{P}(f(x+\varepsilon)=c). \tag{10}
$$

Then $g(x+\delta) = c_A$ *for all* $\|\delta\|_2 < R$, *where*

$$R = \frac{\sigma}{2}\left(\Phi^{-1}\left(\underline{p_A}\right) - \Phi^{-1}\left(\overline{p_B}\right)\right). \tag{11}$$

Theorem 2 demonstrates that Gaussian smoothing inherently leads to ℓ_2 robustness. Specifically, if the assumptions about the base classifier are limited solely to class probabilities as in Eq. (10), then the range of perturbations against which a Gaussian-smoothed classifier can be provably defended aligns precisely with an ℓ_2 ball.

Theorem 2 *[20] Assume* $\underline{p_A} + \overline{p_B} \leq 1$. *For any perturbation* δ *with* $\|\delta\|_2 > R$, *there exists a base classifier* $f(.)$ *consistent with the class probabilities defined in in Eq. (10) for which* $g(x+\delta) \neq c_A$.

Adversarial training and randomized smoothing are two promising directions of defense. The following sections will introduce these two methods and other significant defense methods.

3 Adversarial Attacks

Adversarial attacks represent a significant security threat to artificial intelligence (AI) systems, manifesting primarily during the model prediction phase, as depicted in Fig. 2. This section delves into various adversarial attack methodologies, highlighting their diversity and impact.

According to the degree of knowledge and mastery of modeling, adversarial attacks can be categorized as the generation of white-box adversarial examples and black-box adversarial examples. Figure 5 illustrates the categories of different attacks. In the white-box adversarial attacks, the adversaries can obtain the model knowledge to launch the invasions. In black-box

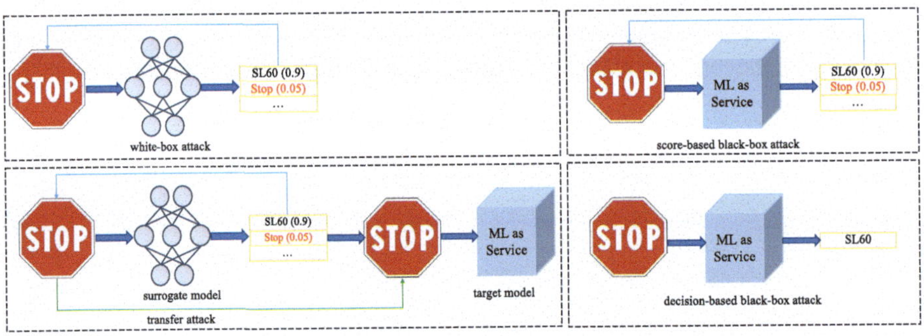

Fig. 5 Taxonomy and illustration of different categories of adversarial attacks

adversarial attacks, the adversaries would carry out a sneak attack on the Machine Learning as a Service (MLaaS) system that the model information is protective to the users. When an attacker obtains a white-box adversarial example with a high attack success rate (ASR) on a source model and migrates it to a target model for an attack, man calls it a black-box transfer attack. The MLaaS systems will provide an output under the black-box adversarial attacks. If the output is a confidence vector, the attack is a score-based adversarial attack. If the output is a specific category, the attack is a decision-based adversarial attack.

Table 2 presents a compilation of seminal digital attack methods developed over the past decade, reflecting the significant scholarly contributions in this domain. The ensuing subsections will elaborate on these methods, providing a comprehensive understanding of their mechanisms and implications.

3.1 White-Box Attacks

The first milestone work of adversarial attack is Limited Memory Broyden-Fletcher-Goldfarb-Shanno (L-BFGS) attack [3], which aims at finding an imperceptible minimum input perturbation in the constraint space of inputs. Equation (12) defines an L-BFGS attack as a box-constrained optimization for the maximization of the loss and the minimization of the perturbation norm:

$$\min_{\delta} c\|\delta\|_2 + \mathrm{Loss}(x + \delta, l), \tag{12}$$

where the symbols x, δ, and l denote the data, the perturbation, and the label. The adversaries perform the line-search mechanism to find the minimum tradeoff parameter c. Goodfellow et al. [4] propose an one-step iterative gradient-based attack method defined in Eq. (6). Basic Iterative Method (BIM or I-FGSM) [21] iteratively solves δ and updates new adversarial samples based on FGSM [4] in multiple steps defined in Eq. (7). The PGD method [6] will project the perturbed input back onto the norm ball. The I-FGSM method combined with the momentum method can evolve the momentum I-FGSM method (MIM) [22]. Equation (13) defines the procedure of the MIM attack, where $\nabla_i J(\cdot)$ is the gradient of the specific time step.

$$\begin{aligned} x_{i+1} &= \mathrm{Clip}\left\{x_i + \varepsilon \cdot \frac{\nabla_{i+1} J(\cdot)}{\|\nabla_{i+1} J(\cdot)\|_2}\right\} \\ \nabla_{i+1} J(\cdot) &= \mu \cdot \nabla_i J(\cdot) + \frac{\nabla_x \mathrm{Loss}\,(x_i, y)}{\|\nabla_x \mathrm{Loss}\,(x_i, y)\|_1} \end{aligned} . \tag{13}$$

The Jacobian Salient Map Attack (JSMA) method [23] can use salient map defined in Eq. (14) learned by the neural networks to generate the adversarial examples.

$$S(\mathbf{X}, t)[i] = \begin{cases} 0 \text{ if } \frac{\partial \mathbf{F}_t(\mathbf{X})}{\partial \mathbf{X}_i} < 0 \text{ or } \sum_{j \neq t} \frac{\partial \mathbf{F}_j(\mathbf{X})}{\partial \mathbf{X}_i} > 0 \\ \left(\frac{\partial \mathbf{F}_t(\mathbf{X})}{\partial \mathbf{X}_i}\right) \left|\sum_{j \neq t} \frac{\partial \mathbf{F}_j(\mathbf{X})}{\partial \mathbf{X}_i}\right| \text{ otherwise} \end{cases}, \tag{14}$$

Table 2 Classical adversarial attack methods. The main text gives the full names of abbreviations

Method	Distance	Physical attack	Knowledge	Iterative	Targeted
L-BFGS [3]	ℓ_2	No	White	Yes	Yes
FGSM [4]	ℓ_∞	No	White	No	No
BIM [21]	ℓ_∞	Yes	White	Yes	No
PGD [6]	ℓ_∞	No	White	Yes	No
MIM [22]	ℓ_∞	No	White	Yes	Both
JSMA [23]	ℓ_0	No	White	Yes	Yes
C&W [5]	$\ell_0, \ell_2, \ell_\infty$	No	White	Yes	Yes
EAD [24]	$\ell_1, \ell_2, \ell_\infty$	No	White	Yes	Yes
EOT [25]	ℓ_2	Yes	White	Yes	Both
BPDA [26]	ℓ_2, ℓ_∞	No	White	Yes	Both
OptMargin [27]	$\ell_0, \ell_2, \ell_\infty$	No	White	Yes	No
AutoAttack [28]	ℓ_2, ℓ_∞	No	White	Yes	Both
DeepFool [29]	ℓ_2	No	White	Yes	No
UAP [30]	ℓ_2, ℓ_∞	No	White	Yes	No
UAN [31]	ℓ_2, ℓ_∞	No	White	Yes	Yes
ATN [32]	ℓ_2	No	White	Yes	Yes
FFF [33]	ℓ_∞	No	White	Yes	No
GD-UAP [34]	ℓ_∞	No	White	Yes	No
ImageNet-C [35]	ℓ_∞	No	Black (Score)	No	No
Perlin [36]	ℓ_∞	No	Black (Score)	No	No
Simplex [37]	ℓ_∞	No	Black (Score)	No	No
Worley [37]	ℓ_∞	No	Black (Score)	No	No
Papernot et al. [38]	ℓ_∞	No	Black (Transfer)	Yes	No
Curls and Whey [39]	ℓ_2	No	Black (Transfer)	Yes	Both
Translation-Invariant Attack [40]	ℓ_∞	No	Black (Transfer)	Yes	No
DI^2-FGSM [41]	ℓ_∞	No	Black (Transfer)	Yes	No
VNI-FGSM [42]	ℓ_∞	No	Black (Transfer)	Yes	No
TREMBA [43]	ℓ_∞	No	Black (Transfer)	Yes	Both
RAP [44]	ℓ_2, ℓ_∞	No	Black (Transfer)	No	Both

(continued)

Table 2 (continued)

Method	Distance	Physical attack	Knowledge	Iterative	Targeted
Yang et al. [45]	ℓ_∞	No	Black (Transfer)	Yes	No
NES Attack [46]	ℓ_2, ℓ_∞	No	Black (Score)	No	Both
$\mathcal{N}$-Attack [47]	ℓ_∞	No	Black (Score)	No	No
AdvFlow [48]	ℓ_∞	No	Black (Score)	No	No
ZO-SignSGD [49]	ℓ_2, ℓ_∞	No	Black (Score)	No	No
Bandit Attack [50]	ℓ_2, ℓ_∞	No	Black (Score)	No	Both
SimBA [51]	$\ell_0, \ell_2, \ell_\infty$	No	Black (Score)	No	Both
ECO Attack [52]	ℓ_2, ℓ_∞	No	Black (Score)	No	Both
Sign Hunter [53]	ℓ_2, ℓ_∞	No	Black (Score)	No	No
Square Attack [54]	ℓ_2, ℓ_∞	No	Black (Score)	No	Both
$\mathcal{CG}$−ATTACK [55]	ℓ_∞	No	Black (Score)	No	Both
Boundary Attack [56]	ℓ_2	No	Black (Decision)	No	Both
OPT [57]	ℓ_2	No	Black (Decision)	No	Both
Sign-OPT [58]	ℓ_2	No	Black (Decision)	No	Both
Evolutionary Attack [59]	ℓ_2	No	Black (Decision)	No	Both
CISA [60]	ℓ_2	No	Black (Decision)	No	No
GeoDA Attack [61]	$\ell_1, \ell_2, \ell_\infty$	No	Black (Decision)	No	Both
HopSkipJumpAttack [62]	ℓ_2, ℓ_∞	No	Black (Decision)	No	Both
QEBA [63]	ℓ_2, ℓ_∞	No	Black (Decision)	No	Both
Sign Flip Attack [64]	ℓ_∞	No	Black (Decision)	No	Both
RayS [65]	ℓ_∞	No	Black (Decision)	No	No

where $\mathbf{F}(.)$ is the salient map denoted by Jacobian matrix, and $S(\mathbf{X}, t)[i]$ is a corresponding saliency map. The adversary modifies the input feature with the saliency map to realize the deceit.

C&W attack [5] tries to find small δ in ℓ_0, ℓ_2, and ℓ_∞ norm. It is an adaptive attack method that the attacker has knowledge of the defense strategies and specifically designs an attack to circumvent or disrupt these defenses. The adaptive attack method is dynamic and purposeful, meaning it adapts to the specific defense mechanisms of the target model. Eq. (15) defines the optimization scheme of the C&W attack.

$$\begin{array}{c} \min_\delta \|\delta\|_p + c \cdot f \mid (x+\delta) \\ f(x+\delta) = \max\left(\max\{Z(x+\delta)_i : i \neq t\} - Z(x+\delta)_t, -\mathcal{K}\right) \end{array}, \tag{15}$$

where c is a hyperparameter, $f()$ is an artificially defined function, and $\mathcal{K}$ is the constraint to assist the generation of adversarial examples. After the proposal of the C&W attack, other variant methods of adaptive attacks have been proposed. The EAD method (Elastic-net Attacks to Deep Neural Networks) [24] transforms the process of attacking Deep Neural Networks (DNNs) using adversarial samples into an optimization problem using elastic-regularized net. OptMargin [27] is another extension of the C&W attack by replacing one objective function with multiple objective functions around the data x. The EOT (Expectation Over Transformation) method is a generalized framework that allows for the construction of adversarial examples that remains the deceptive effect on selected transformation distributions $\mathcal{T}$. The core idea is to constrain the distance between the adversarial input and the original input in the optimization process. Equation (16) defines the defined perturbation of the EOT method, and Eq. (17) describes the formulation of the optimization problem.

$$\delta = \mathbb{E}_{t\sim\mathcal{T}}\left[d\left(t\left(x'\right), t(x)\right)\right], \tag{16}$$

$$\begin{array}{ll} \underset{x'}{\arg\max} & \mathbb{E}_{t\sim\mathcal{T}}\left[\log P\left(y_t \mid t\left(x'\right)\right)\right] \\ \text{subject to} & \mathbb{E}_{t\sim\mathcal{T}}\left[d\left(t\left(x'\right), t(x)\right)\right] < \epsilon \\ & x \in [0,1]^d \end{array}, \tag{17}$$

where $\mathcal{T}$ is the distribution and $t(.)$ means the transformation which is robust to noise, distortion, and affine transformations. In the landmark work on robustness [26], Athalye et al. identified the phenomenon of obfuscated gradients, highlighting false security in certain defense methods under iterative optimization attacks. They also discovered three types of gradient phenomena leading to confusion: Shattered Gradients, Stochastic Gradients, and Vanishing/Exploding Gradients. Moreover, they proposed three attack techniques for inspecting obfuscated gradient types: Backward Pass Differentiable Approximation (BPDA), Expectation Over Transformation (EOT), and Reparameterization [26]. Another notable work is the AutoAttack method [28], which aims to address the misleading impression of robustness by identifying evaluation pitfalls. Based on an open leaderboard, the AutoAttack method can evaluate defense methods for potential gradient obfuscation or masking. It has now become an unwritten rule that defense methods should be assessed on the AutoAttack benchmark.

3.2 Universal Adversarial Perturbations

The aforementioned white-box attack methodologies are tailored to specific models, thereby catalyzing inquiries into model-agnostic assault techniques. Moosavi-Dezfooli et al. [29] computes the minimal adversarial disturbance necessary for a more accurate assessment of robustness. This development has spurred further investigation into universal adversarial perturbations (UAP), capable of affecting a wide range of model architectures. The UAP method [30] identifies perturbations that exhibit transferability across diverse models. Other generative methods including Universal Adversarial Networks (UAN) [31] and Adversarial Transform Networks (ATN) [31] can generate data-specific universal perturbations, while the Fast Feature Fool method (FFF) [33] and Generalizable data-free UAP (GD-UAP) [34] can generate data-independent universal perturbations. In contrast to white-box universal adversarial perturbations, black-box counterparts offer broader applicability in real-world contexts, transforming external security threats into metrics for evaluating internal system safety. ImageNet-C [35] is a benchmark to evaluate the robustness of neural networks to common perturbations. The corruptions include Gaussian noise, shot noise, impulse noise, defocus blur, glass blur, motion blur, zoom blur, snow, frost, fog, brightness, contrast, elastic transform, pixelate, jpeg compression, speckle noise, Gaussian blur, spatter, and saturation. The utilization of procedural noise functions in computer graphics [36, 37] can also generate the textures to deceive the neural networks. The application of ImageNet-C [35], Perlin Noise [36], Simplex Noise [37], Worley Noise [37], and other black-box UAPs can help simulate the adverse weathers and sensor disturbances in the virtual experiment. Such plug-ins are significant for achieving SOTIF in automated driving.

3.3 Black-Box Attacks

Launching the white-box attacks requires knowledge of model structures or data distributions, which limits its scope in the actual applications. In most cases, attackers and defenders do not know each other. Therefore, it is significant to study adversarial robustness under black-box attacks. This sub-section will give a review of the classical black-box adversarial attack methods.

Transfer-based Black-box Attacks

A promising avenue for black-box attack methodologies involves the deployment of transfer-based attacks. Within this framework, adversaries cultivate a surrogate model to mimic the targeted system. It is achieved by employing inputs artificially crafted by the adversary, which are subsequently classified by the target model to predict the wrong labels. Upon mounting a successful attack on the surrogate model, the adversary is then capable of extrapolating the malevolent data to the target model to launch the attack.

In the past decade, many significant studies related to transfer-based black-box attacks have been proposed. Papernot et al. [38] propose the first pioneering work of transfer-based adversarial attack. In this milestone work [38], the attacker generates the adversarial synthetic inputs by a Jacobian-based heuristic and crafts the adversarial examples with a high attack success rate to invade the MLaaS systems. Shi et al. [39] propose a Curls & Whey optimization mechanism to boost the transfer-based attack that the adversaries "curl" up the iterative invasion trajectories to add more diversities and transferabilities in the malicious outputs and further squeeze the "whey" of noise to boost the robustness of perturbations. The white-box adversarial examples would usually be correlated with the discriminative regions of models or gradient trajectories in the optimization process, leading to difficulties in the transferability of adversarial attacks. The tranferability can be improved by data augmentation, including MIM [22] and Diverse Inputs Iterative Fast Gradient Sign Method (DI^2-FGSM) [41]. Wang et al. [42] propose a variant tuning momentum iterative FGSM method (VNI-FGSM) to boost the attack performance. Huang et al. [43] proposed an attack method based on transferable model-based embedding called TRansferable EMbedding-based Black-box Attack (TREMBA). This approach utilizes pre-trained models to learn a low-dimensional embedding space and search within the space to generate adversarial perturbations with high-level semantic patterns to improve the effectiveness of black-box attacks. Yang et al. [45] propose a method to attack the target model via the hierarchical generative networks. Qin et al. [44] propose a method to achieve both targeted and untargeted attacks via the Reverse Adversarial Perturbation (RAP), which finds the stable adversarial examples by minimizing the maximum loss value within a local neighborhood.

Score-based Black-box Attacks

Within a black-box query-based adversarial attack, attackers lack internal model details like weights and structure. Instead, they utilize input and output information to craft effective adversarial examples. One popular methodology is the score-based attack: the attacker adjusts their strategy using the model's score/probability output. The attack typically involves: 1. The attacker makes an exploratory query; 2. The model returns a confidence score; 3. The methods like zeroth-order optimization [66] generate adversarial examples, potentially tricking the model.

The Natural Evolution Strategy (NES) attack methodology, introduced by Ilyas et al. [46], represents a foundational approach to score-based black-box attacks. This pioneering work delineates three distinct real-world scenarios: the query-limited setting, the partial-information setting, and the label-only setting. The NES method adeptly generates black-box adversarial examples within a query-limited context by estimating gradients and constructing adversarial examples through the application of the PGD algorithm on the estimated gradients. In scenarios characterized by partial information, NES strategically perturbs the image by projecting it onto a sphere centered around the original image, thereby maximizing the likelihood of misclassification into the target category while ensuring inclusion within the top k predicted classes. However, the long query time is an obstacle that limits the scalability of the NES method.

Many other studies enhance the paradigm started from the NES attack. Li et al. [47] propose a method called $\mathcal{N}$-Attack to find a probability density distribution in a narrow region centered on the input, from which sampling can increase the success of a black-box attack. AdvFlow [48] is an extension of $\mathcal{N}$-Attack in which the adversary exploits the normalizing flows for constructing the probability density function of adversarial examples. Liu et al. [49] design a zeroth-order stochastic optimization algorithm (ZO-signSGD), which employs the dual advantages of gradient-free operations and the signSGD mechanism to address the problem of black-box attacks. Ilyas et al. [50] form the construction of such a black-box attack as a gradient estimation problem and prove that a least-square estimator is a feasible way to solve this problem. They propose a method based on bandit optimization, enabling the adversaries to integrate priors into the attack settings. Guo et al. [51] propose a simple black-box attack method (SimBA) that utilizes a finite assumption of continuous-valued confidence scores to construct the adversarial images by randomly selecting orthogonal basis vectors and adding or subtracting them in the manipulation process. To augment the efficacy of score-based black-box attacks, Moon et al. [52] propose an efficient combinatorial optimization (ECO) attack method to generate the adversarial perturbations. Further contributing to advancements in this field, Dujaili et al. [53] propose the SignHunter algorithm, which innovatively estimates the sign bit of the gradient during black-box attacks. By leveraging the divisibility characteristic of directional derivatives in the loss function related to the attack, SignHunter employs a partitioning strategy coupled with adaptive querying to ascertain the gradient's sign bit. This method stands out for its remarkable accuracy and efficiency, significantly improving existing techniques. Andriushchenko et al. [54] propose the Square Attack method that utilizes a randomized search scheme, ensuring that the perturbation is strategically crafted near the feasible set of the boundary at each iteration. Applying this black-box method in the object detection task is also successful [67].

Two main challenges remain for the research of the score-based black-box attacks. First, the efficiency problem inherited from the milestone NES attack [46] is still an open issue. Second, some studies [68, 69] demonstrate that these classical black-box methods have difficulties attacking the relatively robust structure like WideResNet [70], Vision Transformer (ViT) [71], and SwinTransformer [72]. The novel score-based black-box attack research in the new paradigm is meaningful.

Decision-based Black-box Attacks

In many real-world applications, the confidence scores of neural network outputs are invisible to users. Instead, the MLaaS systems provide a final decision, not uncertainty information. It emphasizes the importance of decision-based black-box attacks. These are applicable to real AI systems like intelligent vehicles. The decision-based attacks require less knowledge than the transfer-based attacks and are more difficult to defend against than score-based attacks.

The Boundary Attack [56] is the first decision-based attack method in which the adversary starts from a large adversarial perturbation and then seeks to reduce the perturbation while staying the beguiling effect. Cheng et al. [57] postulate that the random-walk method on the boundary, which requires many queries, lacks convergence guarantees. Based on the zeroth order optimization, the OPT method [57] addresses the issue where the decision-based black-box attack is formulated as a real-valued continuous optimization problem. The extension work of Sign-OPT incorporates a direct estimation of the sign of gradient at any direction to the OPT framework [58]. Another extension of Boundary Attack [56] is Customized Iteration and Sampling Attack (CISA) [60] that the adversary estimates the distance based on a dual-direction iterative trajectory from the nearby decision boundary for iterative search of adversarial examples. The Evolutionary Attack [59] method models the local geometry in the search direction and reduces the dimension of the sampling space of adversarial examples. Rahmati et al. [61] propose a geometry-based framework named Geometric Decision-based Attack (GeoDA) to generate black-box adversarial samples where each query returns the highest confidence label of the classifier. The GeoDA framework builds on the assumption that the decision boundaries of neural networks typically have small mean curvature observations in the neighborhood of the data sample. The authors propose an efficient iterative algorithm for generating black-box perturbations with a small p-paradigm ($p \geq 1$), which is validated by the attack experiments on state-of-the-art image classifiers. Chen et al. [62] propose a novel HopSkipJumpAttack method that generates adversarial samples with the Monte Carlo estimation method in a hard-label setting. The algorithm is based on a new gradient direction estimation that uses binary information to estimate the gradient direction on the decision boundary and approximates the optimal solution iteratively. Implementing the zeroth-order gradient estimation in the low-dimensional subspace instead of the original space is a potential query-efficient boundary-based black-box attack (QEBA) method [63]. Chen et al. [64] show that randomly flipping the signs of the entries improves the effectiveness and efficiency of the adversarial attack process. The Ray Searching (RayS) method [65] addresses the inefficiency of decision-based black-box attacks. It builds on the discrete modeling of continuous problems to avoid gradient estimation. Moreover, it eliminates all unnecessary searches through a quick checking step that surprisingly reduces the number of queries required for the attack.

3.4 Physical Attacks

In real-world applications, physical attacks may have more significant impact and research value than digital attacks, especially in the context of autonomous driving. Physical attacks typically encompass three stages: (1) the generation of adversarial perturbations in digital space; (2) the transformation of digital perturbations into physical perturbations with robustness guarantees; and (3) the evaluation of physical perturbations using scanners, cameras, or LiDAR devices. Table 3 lists well-established physical adversarial attack methods, which will be further described in subsequent sections.

Physical attacks should ensure two types of robustness. First, the robustness of digital-to-physical transformation: color space sensitivity can cause physical attack instability. The non-printability score (NPS) metric helps address this issue [100]. Second, the robustness of physical-to-digital transformation: physical adversarial examples should maintain deception under camera distortion, spectral interference, and incomplete echoes in LiDAR and Radar. The adaptive EOT (Expectation Over Transformation) attack method [25] can increase adversarial example robustness across scale or rotation changes. Most white-box physical adversarial attacks follow the NPS-EOT combination paradigm.

Several vision tasks in automated driving would be disturbed by the physical adversarial examples, including traffic sign recognition and detection, traffic recognition, vehicle detection, road line segmentation, monocular depth estimation, and LiDAR perception.

Traffic Sign Recognition and Detection

Eykholt et al. [13] propose a general attack algorithm, Robust Physical Perturbations (RP_2), to generate robust visual adversarial perturbations under different physical conditions. This method can attack the "STOP" sign as the speed-limit sign. This attack paradigm can also be applied in the object detection task to fool the state-of-the-art (SOTA) model [73]. Li et al. [74] propose a novel method in which the adversaries manipulate the translucent sticker over the lens of a camera to fool the traffic sign classifier. Zhao et al. [75] attempt to attack the feature extraction process to boost the physical attack performance. In addressing the Hiding Attack (HA) scenario, they introduce the feature-interference reinforcement (FIR) method alongside the enhanced realistic constraints generation (ERG) approach to bolster robustness. Conversely, for the Appearing Attack (AA), they devise the nested-AE framework, which integrates two autoencoders (AEs) to compromise object detectors effectively at both long and short distances. The patch-based physical attack is easily identified by the human observer, which remains a massive challenge in security research. The Adversarial Camouflage (AdvCam) method [76] is proposed to incorporate the natural style in the physical adversarial examples so that the crypticity of the adversarial examples is increased. Furthermore, the generative adversarial networks (GAN) [101] can be utilized as a data augmentation method to enhance the adversarial attack [77]. Ye et al. [87] apply the adversarial patch method [102] in the attack on the traffic sign recognition. Bai et al. [86] attempt to

Table 3 Classical physical attack methods in automated driving. The main text gives the full names of abbreviations

Method	Knowledge	Tasks
RP_2 [13]	White	Traffic sign recognition
RP_2D [73]	White	Traffic sign detection
CAMOU [14]	Black	Vehicle detection
Adversarial camera sticker [74]	White	Traffic sign recognition
FIR [75]	White	Traffic sign detection
ERG [75]	White	Traffic sign detection
AdvCam [76]	White	Traffic sign recognition
PhysGAN [77]	White	Traffic sign recognition
ER [78]	Black	Vehicle detection
UPC [79]	Black	Vehicle detection
Wu et al. [80]	White	Person detection
Xu et al. [81]	White	Person detection
Boloor et al. [82]	Black	Road line segmentation
Yamanaka et al. [83]	White	Monocular depth estimation
Sun et al. [84]	Black	LiDAR perception
Tu et al. [85]	White	LiDAR perception
IAP [86]	Black	Traffic sign recognition
Adversarial Patch [87]	White	Traffic sign recognition
SLAP [88]	Black	Traffic sign recognition
Adversarial Laser [89]	Black	Traffic sign recognition
Rolling Shutter Effect Attack [90]	Black	Traffic sign recognition
DAS [91]	White	Vehicle detection
Zolfi et al. [92]	White	Traffic light detection
Sato et al. [17]	White	Road line segmentation
Cao et al. [93]	White	Multi-sensor fusion
Shadow Attack [94]	Black	Traffic sign recognition
DTA [95]	White	Vehicle detection
Cheng et al. [96]	White	Monocular depth estimation
Dos Attack [97]	White	Navigation and planning
RP_2-CAM [98]	White	Traffic sign recognition
Cao et al. [99]	White	LiDAR perception

generate the inconspicuous adversarial patches (IAP) to boost the transferability. The IAP method uses the patch generation process in a coarse-to-fine way by utilizing multiple-scale generative models. Lovisotto et al. [88] use a light projector to craft the attacks with the generated Short-Lived adversarial perturbations (SLAP). The laser jamming [89], rolling shutter effect [90], and even shadows [94] can be utilized to craft the adversarial examples to fool the traffic sign classifiers. Adversarial vulnerability can be regarded as the causal confounding effect. Therefore, Yan et al. [98] attack the traffic sign with the guidance of class activation map (CAM) [103] to find the sensitive regions of the targeted attack class. The recent study [98] finds that the attack difficulties increase after the model structures evolved from Convolution Neural Networks (CNNs) [70, 104, 105] to ViTs [71, 72], which raises a new research focus.

Vehicle Detection

The research on attacks on the vehicle detection model has a huge impact on military applications. Zhang et al. [14] propose the first physical vehicle camouflage inspired by both the research of adversarial examples [4] and GANs [101]. Such a milestone method implements a camouflage pattern to hide the vehicle from being detected by state-of-the-art CNN-based object detectors [106, 107]. The proposed method alternates between two threads. First, the attacker trains a neural approximation function to simulate how the simulator applies camouflage to the vehicle and how the vehicle detector performs, given an image of the camouflaged vehicle. Second, the attacker can minimize the approximation detection score by searching for the optimal camouflage. Wu et al. [78] propose an Enlarge-and-Repeat process (ER) method and a Discrete Searching method to generate the adversarial examples fooling the vehicle detectors. The methods effectively produce the mosaic-like adversarial vehicle textures without using the detector's model weights and differential rendering procedure. The limitation of the method is that it is simulated only in the Carla software. Huang et al. [79] propose a Universal Physical Camouflage (UPC) Attack that can fool the region regressors and classifiers simultaneously. Wang et al. [91] propose a Dual Attention Suppression (DAS) method to inhibit model and human attention. Suryanto et al. [95] propose a Differentiable Transformation Attack (DTA) method that the adversary utilizes a Differentiable Transformation Network (DTN) to learn the expected transformations of rendered objects and generate a robust camouflage texture to attack the vehicle detectors with a wide range of transformations.

Person Detection

Another potential security threat in automated driving is the assault aimed at the person detectors. The attackers can craft the adversarial T-shirts, which are robust under different motion gestures [80, 81]. This malicious clothing can play tricks on both surveillance systems and detectors on pedestrians in automated driving. These pieces of research would also be important in the military self-driving applications.

Road Line Segmentation

In automated driving, road line segmentation is a vital task to ensure the vehicles are in the right line locations. However, it is vulnerable to adversarial attacks. Boloor et al. [82] propose a query-based attack that produces a black malicious line to fool the neural networks. To address the problem of camera frame inter-dependencies influenced by vehicle control, Sato et al. [17] formulate the problem with a security-critical attack goal and propose a novel attack method based on the dirty road patches.

Monocular Depth Estimation

Monocular depth estimation refers to estimating the depth information of each pixel in a scene from a single image. It usually involves using computer vision techniques and algorithms to analyze information such as texture, occlusion relationships, and perspective transformations in the image to infer spatial relationships and distances between pixels. It is also sensitive to adversarial attacks. Yamanaka et al. [83] first apply the patch-based attack method in the monocular depth estimation task. Cheng et al. [96] attack the depth estimation model by generating covert object-oriented adversarial patches, and the proposed attack procedure searches the optimization region as well as utilizes the symmetrization methods to deal with the overall contour region to find the most effective attack method. The attack method can attack different target objects and models in real driving scenarios, leading to depth estimation errors and decreased object detection success.

LiDAR Perception and Multi-sensor Fusion

LiDAR sensing plays a vital role in automated driving, which utilizes laser radar (LiDAR) technology to understand the surrounding environment and obtain road information. The LiDAR device can measure the distance from the body to an obstacle by emitting laser light outward. When it encounters an object, the laser light is reflected and received by a complementary metal-oxide semiconductor (CMOS) sensor reflects and receives the laser light. By combining the real-time Global Positioning System (GPS), inertial navigation information, and the calculation of the emission angle, the system can derive the coordinate orientation and distance information of the object in front. The LiDAR device, with powerful information perception and processing capabilities, can sense the road environment and control the vehicle to achieve the intended goal. Sun et al. [84] conduct the first study on adversarial examples of LiDAR perception in automated driving to explore general vulnerabilities in current LiDAR-based perception architectures and find that neglected occlusion patterns in LiDAR point clouds make self-driving cars vulnerable to spoofing attacks. Sun et al. [84]

construct the first black-box spoofing attack based and successfully attack the PointPillars model [108] and the PointRCNN model [109]. However, the proposed method of Sun et al. [84] named LidarAdv only considers the specific frame. Tu et al. [85] propose a method to generate the adversarial mesh which can be placed on a vehicle roof to hide the malicious object and implement the defense experiment under the attacks with the method of data augmentation and Fast Adversarial Training [110]. Cao et al. [99] propose a novel attack method called Physical Removal Attack (PRA), which is capable of selectively removing LiDAR point cloud data of real obstacles by utilizing laser jamming technology, thus causing the obstacle detector of self-driving cars to fail to recognize and locate obstacles, which in turn enables the cars to make dangerous self-driving decisions. Cao et al. [93] design an attack pipeline with non-differentiable cell-level aggregated features to fool both cameras and LiDAR devices with the invisible perturbations. Recently, Wan et al. [97] have investigated a Semantic DoS (Semantic Denial of Service) vulnerability in self-driving planning systems that could lead to unexpected decision-making behaviors in self-driving vehicles, such as sudden braking or abandoning lane changes. This research designs a vulnerability discovery system called PlanFuzz and demonstrates the severity of the vulnerability and possible exploits through case studies of three attack scenarios.

Perspectives of Physical Attacks

Intelligent connected vehicles (ICV) have achieved mass production in the past few years [111]. Many companies have put their self-driving vehicles into the real-world road testing phases. In the research field, the performance metrics on the KITTI dataset [112] tend to be saturated. On the other hand, the neural network models are vulnerable to the attacks. Compared to the digital adversarial examples, the physical adversarial examples would link more to cybersecurity, which is more severe in real applications. The research on physical attacks and their associated defense strategies is still crucial in the future.

4 Defense

This section reviews the classical defense methods proposed in the past few years. One type of defense is empirical defense, which relies heavily on practical experience and intuitive judgment to make a defense against a specific attack. The other type is certified defense, which does not care about the type of adversarial noise but constructs a strict robustness tight lower bound through mathematical or physical modeling. Most defense methods can be categorized under these two categories. Figure 6 describes a coarse-grained category of current defense methods. The empirical defense methods include adversarial training (AT), non-AT training defense, inference defense, and adversarial detection. The certified defense methods can be classified as exact methods or approximate methods. Figure 7 illustrates the different deployment stages of defense methods. In the training stage, adversarial training and other

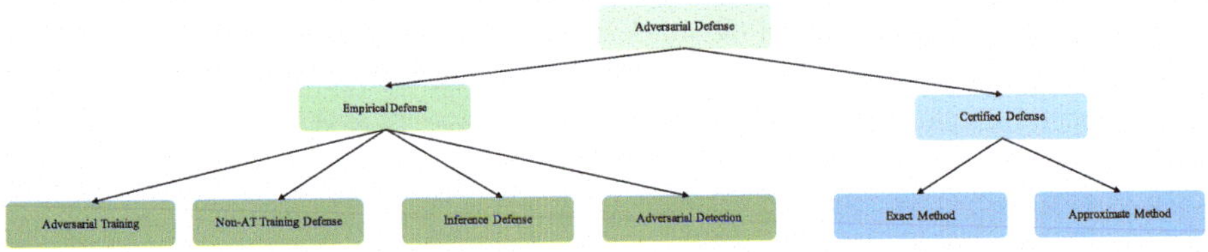

Fig. 6 The category of mainstream adversarial defense method

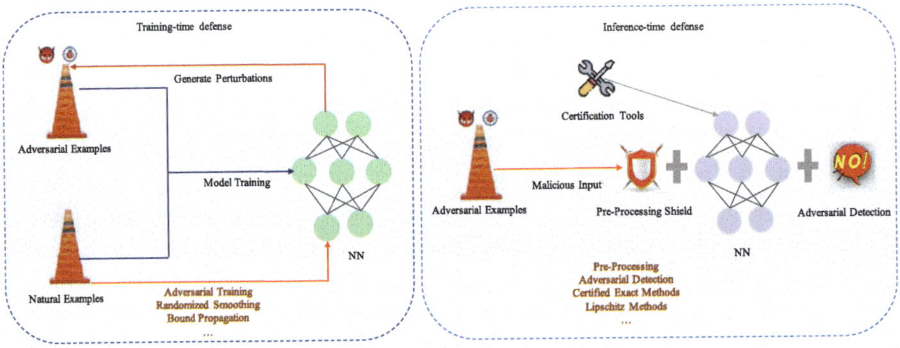

Fig. 7 Different deployment stages of defense methods

approximate certified robustness like randomized smoothing and bound propagation can be applied to train the secure neural networks (NNs). In the inference stage, pre-processing methods can help mitigate the adversarial perturbations, and adversarial detection can reject the malicious input. The certified exact methods and Lipschitz methods can be utilized as the analysis tools to provide a robustness bound.

4.1 Empirical Defense

Empirical defenses improve the robustness of the models against specific adversarial sample attacks through specific methods. Such defenses include adversarial training [6], modification of model structure [68], de-noising [154], randomization [137], and other methods to guarantee model robustness. The key advantage of empirical defenses is that they are typically easy to implement and can provide effective defense mechanisms against various attacks. However, the main disadvantages of these approaches are that they usually rely on specific types of attacks and, therefore, may need to be robust enough against unknown attacks or slightly changing forms of attacks. Table 4 lists the classical empirical defense methods in the past decade.

Table 4 Classical empirical defense methods categorized by addressed problems. The main text provides the full names of abbreviations

Method	Robust generalization	
	Defending attack types	Adversarial training?
Defense Distillation [113]	White	No
Vanilla Adversarial Training [6]	White	Yes
Ensemble Adversarial Training [114]	White	Yes
Deep Defense [115]	White	No
ATDA [116]	White	Yes
TRADES [7]	White	Yes
MART [117]	White	Yes
CCAT [118]	White	Yes
Friendly Adversarial Training [119]	White	Yes
DVERGE [120]	White	Yes
Bag of Tricks for AT [121]	White	Yes
Adversarial Wavelet Training [68]	White	Yes
DM-Improves-AT [122]	White	Yes
GeodesicAT [123]	White	Yes
Method	*Adversarial detection*	
	Defending attack types	Adversarial training?
Metzen et al. [124]	White	No
SafetyNet [125]	White	No
MagNet [126]	White	No
GMM [127]	White	No
Mahalanobis distance [128]	White	No
Reverse Cross Entropy [129]	White	No
CD-VAE [130]	White	No
Libre [131]	White	No
Blacklight [132]	Black	No
PRADA [133]	Black	No
SD [134]	Black	No

(continued)

Table 4 (continued)

	Inference-time defense	
Method	Defending attack types	Adversarial training?
Input transformations [135]	White	No
PixelDefend [136]	White	No
Randomization [137]	White	No
BaRT [138]	White	No
Mixup Inference [139]	White	No
RND [140]	Black	No
DiffPure [141]	White	No
AAA [142]	Black	No
Boundary defense [143]	Black	No
Anti-adversaries [144]	White	No
Dent [145]	Black	No
EBM+DSM [146]	White	No
SOAP [147]	White	No
	Training efficiency	
Method	Defending attack types	Adversarial training?
Free Adversarial Training [148]	White	Yes
YOPO [149]	White	Yes
Fast Adversarial Training [110]	White	Yes
Local Linearity Regularizer [150]	White	Yes
GradAlign [151]	White	Yes
FrequencyLowCut Pooling [152]	White	Yes
Robust critical fine-tuning [153]	White	Yes

Adversarial Training

By far, adversarial training is the most effective way to conduct empirical defense. Such a mechanism can realize data augmentation of adversarial examples in the training process. Moreover, the min-max optimization builds the maximum perturbations and the minimum empirical risks [6] to realize the "gradient penalty" paradigm, which is beneficial to defend against the gradient-based attacks.

Madry et al. [6] propose the first milestone work in adversarial training. However, two open issues are obvious. First, there exists degradation for clean accuracy, while the robust accuracy is not considerable. Another issue is training efficiency that it takes almost two days to train a robust adversarially-trained WideResnet on the CIFAR dataset [155]. Therefore, most sequential studies of vanilla adversarial training [6] focus on either boosting robust generalization or improving training efficiency.

Robust generalization improvement is still a valuable research problem in adversarial training. The Vanilla Adversarial Training method [6] utilizes the white-box PGD method as the default attack strategy. It does not pay attention to the defenses against transfer-based black-box attacks. It may converge to a degenerate global minimum, where small curvature artifacts corrupt the data point and obfuscate the decision of the neural networks. To address this issue, Tramèr et al. [114] conduct an Ensemble Adversarial Training method that augments the training data with perturbations transferred from other models. This method performs considerably on both CIFAR-10 dataset [155] and ImageNet dataset [156]. Adversarial Diversity Promoting (ADP) [157] and DVERGE [120] are preeminent successors of this empirical defense method. Adversarial examples can also be regarded as the distribution with the divergence using the distribution of natural examples as a reference. Song et al. [116] proposes a paradigm of adversarial training with domain adaptation (ADPA) to boost the robust generalization of neural networks. Since the adversarial perturbations can be regarded as the abnormal noises [12], the classical non-local denoising method [158] can be incorporated in adversarial training to formulate the Feature-Denoising Adversarial Training (FD-AT) framework [154]. Another similar method is ME-Net (Matrix Estimation Net) in which the adversarially-trained neural network leverages matrix estimation (ME) to reconstruct images and mitigate perturbations [159].

The milestone work to improve robust generalization is TRADES [7]. Zhang et al. [7] seek a tradeoff between robustness and accuracy with the proof of a tight differentiable upper bound using the theory of classification-calibrated empirical risks. The TRDES method divides the error against robustness into two components, the estimation error against natural samples and the boundary error, and estimates upper bounds for each of them. In this case, the upper bound estimation for the natural error uses a convex loss function. The upper bound estimation for the boundary error uses a geometric metric based on the loss function, such as KL Divergence.This approach captures the tradeoff between robustness and accuracy of the model well and provides theoretical guarantees. Yan et al. [123] give proof that the geodesic is the shortest trajectory between two points and propose the Geodesic Adversarial Training (GeodesicAT) framework to enhance the TRADES method. Besides the min-max optimization mechanism, data quantity and quality would also decide the robustness of neural networks. Alayrac et al. [160] postulate that the unlabeled data significantly improves robustness and propose an Unsupervised Adversarial Training (UAT) method to deploy a robust machine learning model. The Misclassification Aware Adversarial Training (MART) method [117] is another milestone work that the correctly-classified/incorrectly-classified training samples are regularized in different ways during the adversarial training process. Balunovic et al. [161] propose a Convex Layerwise Adversarial Training (COLT) method to bridge the gap between adversarial training and provable defense. Chan et al. [162] propose the Jacobian Adversarial Regularized Network (JARN) method with the utilization of optimizing the saliency of a classifier's Jacobian by adversarially regularizing the model's Jacobian to resemble natural training images. Then, the method is extended to the frequency domain [163] to boost adversarial robustness. Adversarial robustness improvement can also connect with the uncertainty calibration to formulate Confidence-calibrated Adversarial Training (CCAT) [118] or return to the tradition of cybernetics to build a robust Close- Loop Control Neural Network (CLC-NN) [164]. The original adversarial training method [6] suffers from the phenomenon of "robust overfitting" [165]. Many sequential studies have been proposed to alleviate such an issue. Zhang et al. [119] prove that the fixed large attack step size may lead the neural network to be immersed in the local optima of robustness and propose a Friendly Adversarial Training (FAT) method based on the curriculum learning mechanism. Furthermore, the Geometry-aware Instance-reweighted Adversarial Training (GAIRAT) method [166] adaptively assigns the larger weights to the difficult adversarial examples. Improving the sample efficiency of adversarial examples is another feasible direction in which man can generate the adversarial distributions rather than the point-wise adversarial examples [167]. The regularization method is also important in adversarial training, e.g., weight decay and early stopping mentioned in the study of "Bag of Tricks for AT" [121], adversarial weight perturbation (AWP) [168], data augmentation [169]. Recently, Pang et al. [170] advocate for the employment of local equivariance as a means to delineate the ideal behavior of a robust model. This approach leads to the formulation of a self-consistent robust error, which they have named SCORE. The new incremental work of adversarial training is boosting the network representation through the wavelet regularization [68], Diffusion-Model-Improves-Adversarial-Training (DM-Improves-AT) [122], "Learnable Attack Strategy" Adversarial Training (LAS-AT) [171] based on the REINFORCE algorithm [172]. In summary, robust generalization improvement is still a core problem of adversarial training. The gap between robust and natural accuracy in a large-scale dataset like ImageNet [156] is still huge and deserves further study.

Besides robust generalization, training efficiency is another issue in adversarial training research. Shafahi et al. [148] first attempt to address this problem to eliminate the overhead of adversarial perturbation generation. The Free Adversarial Training method recycles the gradient information during optimization, which can realize adversarial robustness on the ImageNet dataset in a single workstation with 4 P100 Graph Processing Units (GPUs). The training time is only two days. Zhang et al. [149] prove that adversarial training can be regarded as a discrete-time differential game. Based on Pontryagin Maximum Principle (PMP), they have proposed a You Only Propagate Once (YOPO) method, in which the forward and backpropagation can only be restricted within the first layer of the neural network during the model parameter update process. Wong et al. [110] show that it is possible to train empirically robust models with the FGSM method. It further reduces the training time. Qin et al. [150] show that promoting linearity can alleviate the gradient obfuscation problem of adversarial training and accelerate the training speed. Improving fast adversarial training is a fascinating research direction. Andriushchenko et al. [151] propose a new GradAlign regularization method to alleviate the "catastrophic overfitting" issue by maximizing the gradient alignment during the attack process. Grabinski et al. [152] postulate that poor downsampling operations cause aliasing artifacts and contribute to the adversarial vulnerability of neural networks. Therefore, the proposed FrequencyLowCut pooling method can be combined with the fast FGSM adversarial training method to improve the training efficiency of adversarial defense. The new robust critical fine-tuning method [153] can enhance robust generalization in non-robust critical modules with light training costs. Overall, the reduction of adversarial training costs is still an open problem.

Currently, on large vision datasets like ImageNet [156] and Cityscapes [173], the gap between robustness and clean accuracy is still huge. It highlights the need for continual research on adversarial training. In the cyber-physical system of automated driving, a fast adversarial training method can ensure both the security and safety of vehicles, which deserves further study.

Other Training-stage Defense Methods against Adversarial Examples

Besides adversarial training, other training-stage defense methods are still feasible to improve adversarial robustness.

Yan et al. [115] propose a Deep Defense method with the introduction of an adversarial perturbation-based regularization item in the loss function. The ADP method [157] would encourage the diversity of the decision output based on the ensemble mechanism. The stable neural ordinary differential equation (ODE) model [174] is also important to defend against adversarial attacks. The Defense Distillation method [113] can provide a shield under gradient-based attacks, while it is vulnerable under adaptive attacks.

Overall, the adversarial training methods are the preferred choices for adversarial defenses. However, other techniques like the distillation method [113] and neural ODE [174] are still encouraging. The distillation methods have been utilized in the perception model development of automated driving [175, 176]. Although the distillation method is weak under adaptive attacks, it can be a feasible method against UAPs and other black-box noises.

Inference-time Defense

Sometimes, the adversaries launch the attack off-guard. The deployment of an adversarially-trained model would be too late to provide the shield. The inference-time defense can play as an expedient plug-in in the AI software infrastructure. The input transformation is an intuitive defense method [135], including bit-depth reduction, JPEG compression, total variance minimization, and image quilting. Utilizing the Randomization method [137] can also provide the elastic defense which confuses the adversaries. Raff et al. [138] explore a similar idea of a stochastic Barrage of Random Transformations (BaRT) to defend against adaptive attacks. . One assumption exists that adversarial examples are mainly present in the low probability region of the training distribution. Therefore, Song et al. [136] propose a new method based on generative models, PixelDefend, using statistical hypothesis testing and pixel purification to defend against attacks, building a robust neural network model. Pang et al. [139] propose a method known as Mixup Inference, which involves blending the input with other random, clean samples. This technique is designed to shrink and transfer the equivalent perturbation if the input is adversarial. Self-supervised Online Adversarial Purification (SOAP) [147] is also a novel defense strategy that utilizes the label-independent nature of self-supervised signal to mitigate adversarial perturbations. Building upon the observation that adversaries are generated through iterative minimization of a network's prediction confidence, Alfarra et al. [144] design an anti-adversary method to prevent the construction of adversarial examples. Furthermore, since the diffusion model can be utilized for denoising, Nie et al. [141] propose an adversarial purification method based on the diffusion model [177].

The black-box attacks would be more common in the real applications. Qin et al. [140] propose a lightweight defense method of random noise defense (RND) for the score-based black box attacks, which adds random noise to each query to interfere with attackers' gradient estimation or random search, thus reducing the attack efficiency. Chen et al. [142] propose an Adversarial Attack on Attackers (AAA) method to fool the greedy attackers into incorrect directions by slight perturbations on the neural network outputs in the test time. This method has three advantages: (1) the mitigation of the score-based black-box attacks, (2) the preservation of clean accuracy, and (3) uncertainty calibration. Wang et al. [145] leverage the defensive entropy minimization (dent) mechanism to output the robust prediction under the white-box, black-box, adaptive attacks on CIFAR-10/100 and ImageNet dataset. The recent Boundary Defense method [143] can also guard the MLaaS system, that the model will detect the boundary samples as those with low classification confidence and add white Gaussian noise to their logits.

Compared to the training-time defense methods, the inference-time defense needs less computation costs. It is easy to deploy in real applications such as automated driving and unmanned aerial vehicles.

Adversarial Detection

Adversarial examples can be viewed as anomalous data. The straightforward defense method is to detect them. One questionable view is that the adversarial perturbations are usually not perceptible, and some attacks based on ℓ_0-norm and ℓ_2-norm will limit the changes to pixels. Nevertheless, this type of defense method cannot be neglected.

In the early years of adversarial examples research, adversarial detection is a popular method to defend adversarial examples. The binary classifier [124], the SafetyNet based on the Support Vector Machines (SVM) [125], MagNet with diverse separate detector networks and a reformer network based on the manifold assumption [126] show their considerable performance on adversarial detection. Another detection tools include Gausian Mixture Model (GMM) [127], Mahalanobis distance [128], reverse cross entropy [129], and local intrinsic property [178]. The generative model can also be leveraged to detect adversarial examples. For example, Yang et al. [130] build a class-disentanglement variation autoencoder (CD-VAE) to detect adversarial examples.

The detection of black-box adversarial examples has received huge attention in recent years. PRADA [133] is the first detection model to defend the transfer-based black-box attacks. The stateful detection (SD) method [134] assumes that the attack query sequence exhibits high similarity due to the iterative attack search. The MLaaS system can reject the query and ban the malicious account. The Blacklight method [132] inherits such an assumption and replaces the ℓ_2 distance metric utilized in the SD work [134] with the fingerprints.

The detection methods have demonstrated superior performance in the black-box defense [132, 134]. Whether deploying it into automated driving is worthwhile is still under scrutiny.

4.2 Certified Defense

The empirical defense method would meet the challenges of sophisticated adaptive attackers [26, 223]. The continuous arms between attackers and defenders motivate a theoretical interpretation of adversarial robustness. The certified defense methods respond to these commands. It consists of a robustness verification approach providing the lower bound of robustness under any attacker without the specification of perturbation type and corresponding robust training methods. This subsection reviews the certified defense methods. Table 5 lists the classical certified defense methods in the field of robustness research.

Formal Verification Methods

The formal verification methods are used to formally verify the robustness of a model to specific input perturbations through solvers or theorem-proving techniques. It is an exact but computationally expensive method. Katz et al. [179] focus on the non-convex Rectified Linear Unit (ReLU) activation function, an important ingredient in CNNs. The scalable simplex method of linear programming named Reluplex supports the ReLU constraint. The cost of exactness is a high computation budget, which limits the usage of Reluplex in real applications. Huang et al. [180] utilize the Satisfiability Modulo Theory (SMT) to provide a verification framework on several datasets, including MNIST [224], CIFAR10 [155], GTSRB [225], and ImageNet [156]. Ehlers et al. [181] provide the global linear bound for the piece-wise feed-forward neural networks and reduce the computation cost of SMT on the small dataset. Overall, the formal verification method can provide a relatively exact bound. However, these methods suffer from the high computation cost, which is not applicable for large and foundation models [226].

Table 5 Classical certified defense methods. The main text gives the full names of abbreviations

Method	Degree	Category	Large datasets?
Reluplex [179]	Exact	Formal verification	No
Huang et al. [180]	Exact	Formal verification	Yes
Ehlers et al. [181]	Exact	Formal verification	No
Cheng et al. [182]	Exact	Mixed integer programming	No
Cross-Lipschitz regularization [183]	Approximate	Lipschitz continuity	No
Xiang et al. [184]	Exact	Mixed integer programming	No
Branch-and-Bound [185]	Exact	Mixed integer programming	No
Convex Outer Adversarial Polytope [186]	Approximate	Convex relaxation	No
Random Projection [186]	Approximate	Convex relaxation	No
Dvijotham et al. [187]	Approximate	Convex relaxation	No
Semi-definite Programming [188]	Approximate	Convex relaxation	No
ReLUVal [189, 190]	Approximate	Bound propagation	No
Fast-Lip [191]	Approximate	Lipschitz continuity	No
LMT [192]	Approximate	Lipschitz continuity	No
Richards et al. [193]	Approximate	Cybernetics	No
Fazlyab et al. [194]	Approximate	Lipschitz continuity	No
GeoCert [195]	Approximate	Convex relaxation	No
Salman et al. [196]	Approximate	Convex relaxation	No
IBP [197]	Approximate	Bound propagation	Yes
Lipschitz norm ball [198, 199]	Approximate	Lipschitz continuity	No
PixelDP [200]	Approximate	Randomized smoothing	Yes
Vanilla Randomized Smoothing [20]	Approximate	Randomized smoothing	Yes
Salman et al. [201]	Approximate	Randomized smoothing	Yes
Mangal et al. [202]	Approximate	Uncertainty	No
PROVEN [203]	Approximate	Uncertainty	No
Fazlyab et al. [204]	Approximate	Uncertainty	No
Wang et al. [205]	Approximate	Cybernetics	No
Wang et al. [206]	Approximate	Cybernetics	No
Chiang et al. [207]	Approximate	Bound propagation	No
Sparse polynomial optimization [208]	Approximate	Lipschitz continuity	No
Jordan et al. [209]	Approximate	Lipschitz continuity	No
F-Divergence Smooth [210]	Approximate	Randomized smoothing	Yes
ℓ_∞-distance [211, 212]	Approximate	Lipschitz continuity	No
PointGuard [213]	Approximate	Randomized smoothing	No
GCP-CROWN [214]	Approximate	Bound propagation	No
SortNet [215]	Approximate	Lipschitz continuity	Yes
Schuchardt et al. [216]	Approximate	Randomized smoothing	Yes
Li et al. [217]	Approximate	Randomized smoothing	Yes
LipsFormer [218]	Approximate	Lipschitz continuity	Yes
3deformrs [216]	Approximate	Randomized smoothing	Yes
Alfarra et al. [219]	Approximate	Randomized smoothing	Yes
Anderson et al. [220, 221]	Approximate	Randomized smoothing	Yes
Pfrommer et al. [222]	Approximate	Randomized smoothing	Yes

Mixed Integer Programming Methods

The utilization of mixed integer programming methods also provides a potential solution to the robust verification of neural networks. Cheng et al. [182] quantize the maximum perturbation bound via the mixed integer programming (MIP) method and apply the method in the agent game for safety-critical applications like automated driving. The reachability analysis can be built for both feed-forward ReLU neural networks [227] and Multi-Layer Perception (MLP) [184]. The Branch-and-Bound method can also be used to build a unified view of verification on the small neural networks [185]. None of these methods have been applied to the large-scale scenarios.

Convex Relaxation Methods

Although most neural networks handle non-convex optimization problems, the convex relaxation methods can be utilized in the certified defense framework. Compared to exact methods, convex relaxation methods are approximate numerical optimization techniques for solving convex problems. Convex relaxation methods find the global optimal solution by gradually relaxing the original constraints of the problem into a series of approximate convex optimization sub-problems. Wong et al. [186] utilize a convex outer approximation of the reachable sets. It is shown that the dual problem of convex outer adversarial polytope can represent the backpropagation of DNNs. The Random Projection method [186] scales the provable defense method to the complex scenario. Inspired by the idea of duality, Dvijotham et al. [187] formalize the problem of certified defense as an unconstrained convex optimization problem and obtain a provable robust boundary by solving a Lagrangian relaxation of this unconstrained convex optimization problem. Sub-gradient methods can solve this computing process of the robust boundary. Raghunathan et al. [188] propose a Semi-definite Relaxation method to solve the max-cut problem for a certified robustness bound. The sequential improvement work of convex relaxation methods can be built on the polyhedral complices [195]. Furthermore, Salman et al. [196] propose a tight convex relaxation barrier method in a hierarchical framework. Briefly, the convex relaxation methods are approximation methods with high efficiency. However, the scale of these methods on the large-scale datasets is difficult.

Bound Propagation Methods

The bound propagation method is a conservative approximation method to simplify the robust verification process with the computation by calculating the interval range of the output over the input at each layer. Gowal et al. [197] propose a method called Interval Bound Propagation (IBP) to compute the worst-case bounds of the network output by propagating upper and lower bounds on the activation values at each layer. By optimizing the network parameters, the bounds of the network output are made to satisfy the given specification.

The ReLUVal method [189, 190] using symbolic intervals can provide a tight formal security and safety bound for neural networks. Chiang et al. [207] abstract a certified defense model and resolve the problem with the IBP method. Zhang et al. [214] generalize the bound propagation method with the general cutting plane (GPC) to realize robust verification in the GCP-CROWN framework.

Lipschitz Continuity Methods

Lipschitz continuity is a concept of stability analysis. Specifically, a function is said to be Lipschitz continuous on a domain if there exists a real constant $L \geq 0$, known as a Lipschitz constant, such that for the data x_1, x_2, the function output change satisfies the inequality:

$$|f(x_1) - f(x_2)| \leq L |x_1 - x_2|, \tag{18}$$

for all x_1 and x_2 in the domain f.

Hein et al. [183] firstly scale the Lipschitz continuity concept to the certified robustness problem and propose a Cross-Lipschitz regularization method for the defense. The Lipschitz continuity tools can be built on the spheres with norm balls to provide a formal robustness guarantee that does not depend on the space size [198, 199]. Weng et al. [191] propose a fast Lipschitz (Fast-Lip) method on the MNIST [224] and CIFAR [155] datasets with the speed acceleration. Tsuzuku et al. [192] propose a training method called Lipschitz-Margin Training (LMT), which improves the certified robustness of the neural networks by calculating an upper bound on the Lipschitz constant of each component and using that upper bound to train the network with looser robust bounds. Fazylab et al. [194] show the activation functions can be interpreted as gradients of convex potential functions and calculate the Lipschitz constant with the semi-definite programming method. Jordan et al. [209] calculate the non-smooth vector-valued functions via the norm of the generalized Jacobian. Latorre et al. [208] implement the Lipschitz constant estimation for certified robustness via the sparse polynomial optimization mechanism. The Lipschitz constant can also be calculated via the ℓ_∞-distance [211, 212] or the boolean function [215]. Qi et al. [218] scale the Lipschitz continuity method to the vision transformer model (LipsFormer) and replace the unstable LayerNorm model with the Lipschitz continuous CenterNorm module.

Generally speaking, the Lipschitz continuity methods can handle the neural networks with a non-differentiable input transformation, which is suitable for analyzing the activation function in deep learning. However, the Lipschitz continuity methods would be inclined to output the looser robust bound.

Randomized Smoothing Methods

The randomized smoothing methods improve the robustness of the model to perturbations by adding random noises around the input data and averaging multiple noisy perturbations of the model. These approaches provide probabilistic robustness guarantees.

Lecuyer et al. [200] propose the Pixel Differential Privacy (PixelDP) method, a novel provable defense mechanism against adversarial examples attacks in a specific range. The method is based on the concept of differential privacy, which can be applied to any type of deep neural network and can be applied to large-scale networks and datasets. The main idea of PixelDP is to add noise to the training and prediction time to increase robustness while maintaining provable privacy protection. The Neyman-Pearson lemma [228] provides a relatively accurate boundary for binary hypothesis testing (robust or vulnerable). Cohen et al. [20] propose a robust certified defense method under the ℓ_∞-norm attacks based on the Neyman-Pearson lemma. Besides, Cohen et al. [20] also show that the Monte Carlo algorithm can evaluate the prediction trustworthiness of smooth classifiers. Salman et al. [201] design an adaptive attack mechanism for the randomized smooth classifier, which provides a robustness guarantee under strong attacks. Dvijotham et al. [210] prove the robustness of smoothed classifiers via the tools of F-Divergence. Randomized smoothing would cause several hidden costs that shrink the decision boundaries with the adoption of the prediction rules [229]. Moreover, the augmented perturbations do not necessarily solve the boundary shrinkage problem. However, it can help the application of these methods on large-scale datasets. The surprising utilization is on the point-cloud datasets including PointGuard [213], invariance-aware randomized smoothing certificate [216], 3Deform Randomized Smoothing (3deformrs) [230]. Other incremental studies of randomized smoothing include double sampling randomized smoothing [217], projected randomized smoothing [222], data-dependent randomized smoothing [219], optimal randomized smoothing via the semi-infinite linear programming [220], locally-biased randomized smoothing [221].

In summary, the randomized smoothing methods are the acknowledged feasible methods that can be adapted to large-scale datasets. Nevertheless, the shrinking boundary accompanied by smoothing is an open issue. There is still ample room for further research and development in the field of random smoothing.

Uncertainty-based Methods

Certified robustness can connect with uncertainty quantification. Mangal et al. [202] introduce a novel concept of robustness termed probabilistic robustness, necessitating that the neural network exhibits robustness with a probability of at least 1-ε concerning the input distribution. This probabilistic approach is pragmatic and offers a systematic method for assessing the robustness of a neural network. The PROVEN method [203] realizes probabilistic robustness in the case of adversarial perturbations that follow a specific probability distribution, providing probabilistic guarantees that the top-1 predictions of the model will not change in a statistically significant sense of verifiable robustness. Fazlyab et al. [204] compute a confidence ellipsoid for the output via semi-definite programming. The uncertainty-based method is intuitive, and can provide a loose robustness bound. However, its assumption of the noise distribution is not always satisfactory in real applications.

Cybernetics-based Methods

Modern AI methods can also return to the cybernetics tradition. Therefore, the certified defense methods can also be combined with the robust control framework [205], the optimal transport algorithm built on the Feynman-Kac Formalism [206], and the Lyapunov function [193]. However, the effectiveness of these methods needs further inspection.

5 Adversarial Examples in Perception Systems of Automated Driving

The existence of adversarial examples poses a security threat to autonomous driving. This section reviews the significant progress of adversarial examples in automated driving.

5.1 Objection Detection

Object detection is an important task in automated driving. The object detectors based on CNNs [231] and transformers [232] can handle most scenarios in automated driving. However, due to the uncertainty of prediction and potential security issues, it is still far away from trustworthiness. For example, many studies show that object detectors are vulnerable to adversarial examples.

Xie et al. [233] show that both segmentation and detection models will classify multiple objects in an image. Therefore, the attacks can aim at pixels and proposals. Based on the assumption, the Dense Adversary Generation (DAG) method [233] attacks the object detection and semantic segmentation models. Wei et al. [234] propose a method to manipulate the feature maps extracted and improve the transferability on the adversarial examples of object detection. Adversaries can manipulate the patch to defraud the object detectors [235–237]. Huang et al. [238] propose a novel single-model-based black-box adversarial attack method to improve the transferability of attacks against the object detection models. The method is mainly based on a self-ensemble strategy, which includes integrating input data, an attacked model, and an adversarial patch to enhance the transferability of the adversarial patch. Several methods have been proposed to defend against the attacks, including the multi-task learning [239], class-aware robust optimization [240], and adversarially-aware convolution module to disentangle gradients for optimization on clean and adversarial data [241]. Achieving adversarial robustness for object detection in automated driving is still a long way to go.

5.2 Semantic Segmentation

In automated driving, semantic segmentation is a critical task that involves correlating each pixel in an image to a specific category label. This process aims to divide the image into areas with a clear semantic meaning to help autonomous driving systems understand and parse the road environment. Through semantic segmentation, automated vehicles can accurately identify various elements of roads, pedestrians, vehicles, and traffic signs and assign them the corresponding labels. In this way, vehicles can make decisions based on the label information, such as avoiding pedestrians and vehicles, obeying traffic rules, etc. Currently, the recognition performance of the semantic segmentation model is satisfactory [242–244]. However, the adversarial attacks will cause the degradation of the Mean Intersection over Union (mIoU).

After the proposal of DAG [233], the vulnerability problem of dense pixel classification in segmentation still needs to be addressed. Gu et al. [245] propose a segmentation-specific PGD called SegPGD. The adversarial training mechanism based on SegPGD will boost the adversarial robustness of the semantic segmentation model. Yin et al. [246] give a systematic evaluation of adversarial robustness for CNN-based semantic segmentation models in automated driving. The ViT-based semantic segmentation models have become the mainstream [243, 244] in automated driving. The robustness study of these models is still an open problem.

5.3 Adversarial Examples in 3D Perception

LiDAR (Light Detection and Ranging) technology plays a pivotal role in the advancement of automated driving systems, offering superior capabilities in obstacle detection, localization, and navigation compared to traditional camera sensors. LiDAR devices boast higher resolution ratios for distance, angle, and speed, coupled with robust anti-interference properties, making them particularly effective under adverse weather conditions. Many contemporary commercial autonomous vehicles leverage systems that integrate LiDAR with camera devices for enhanced perception [247]. Architectures relying solely on LiDAR [108, 248, 249] as well as those employing sensor fusion techniques [250, 251] have demonstrated significant achievements on academic benchmarks such as KITTI [112] and nuScenes [252].

Despite these advancements, security research has illuminated vulnerabilities within LiDAR-based systems, revealing potential for spoofed attacks and the generation of malicious obstacles [16, 84, 85, 253]. These adversarial challenges extend to multi-sensor fusion models as well, exposing similar susceptibilities [93, 254]. In response, several exhaustive investigations have endeavored to assess the robustness of LiDAR-based 3D object detection [255, 256] and sensor fusion models [9]. However, as the landscape of foundational and planning-oriented models evolves [257, 258], the focused exploration into their adversarial robustness becomes increasingly significant.

5.4 Trajectory Prediction

Trajectory prediction stands as a pivotal component in the applications of automated driving, tasked with forecasting the movements of nearby vehicles and pedestrians to prompt the control, planning, and navigation strategies. While these models have demonstrated impressive efficacy in naturalistic settings [259–261], their susceptibility to adversarial attacks poses a severe challenge [15, 262, 263]. In response, diverse defensive strategies have been advanced, such as domain-specific data augmentation and adversarial training [264], semi-supervised semantics-guided adversarial training, and adversarial defenses that leverage causal Total Direct Effect (TDE) inference [265]. Currently, many pieces of research still focus on the white-box attack scenarios. However, most automated driving systems are black-box MLaaS systems. Therefore, the research community needs to allocate more attention to the black-box adversarial robustness of trajectory prediction models.

6 Adversarial Examples and SOTIF

Driving safety, traffic efficiency, and low-carbon transportation are several significant factors for automated vehicles. The driving safety includes functional safety, cybersecurity, and SOTIF. Usually, the research of adversarial examples belongs to the category of cybersecurity. Currently, there are three critical challenges in SOTIF [247]: the long-tailed scenario problem, the system complexity and diversity in automated driving, and the AI algorithm's inexplicability and uncertainty. All these issues connect with the adversarial robustness of automated vehicles. Firstly, the existence of adversarial obstacles and malicious traffic signs plays a critical role in the operational environment of self-driving cars and thus cannot be overlooked. Secondly, while overparameterization has been shown to enhance the adversarial robustness of neural networks [266], there is a pressing need for practical systems to adopt lightweight yet robust models [267]. Thirdly, the relationship between the uncertainty inherent in AI methodologies and adversarial robustness has been the subject of several studies [268–270], but all these studies have not been extended to automated driving scenarios. The offline safety design, online safety monitoring, and active ongoing learning [247] should take adversarial robustness into consideration. Moreover, the universal adversarial examples [35, 37] offers a unique opportunity to simulate adverse weather conditions and sensor failures in automated driving, serving as practical test cases for AI system evaluation and enhancement through active learning. By reinterpreting external security threats as catalysts for improving internal system safety, we can shift the paradigm towards a more resilient automated driving ecosystem. Despite their significance, these areas of study have yet to garner the attention they merit. Moving forward, they represent critical avenues for in-depth exploration by the research community.

7 Future Research Directions

There are several research directions related to adversarial robustness in automated driving.

Firstly, the landscape of foundational models in computer vision has witnessed significant advancements in recent years. Among these, the Segment-Anything model [271] demonstrates the capability for zero-shot recognition across various application domains, though its efficacy and robustness within the context of automated driving remain to be fully explored. Furthermore, recent innovations have introduced a novel architectural framework that facilitates sequential modeling over linear time through the utilization of selective state spaces [272, 273], effectively addressing the computational challenges associated with processing long sequences by Transformers. This development heralds the emergence of a new avenue for research.

Secondly, the online automated driving algorithms need to run on specific chips rather than NVIDIA A100 GPUs. Therefore, addressing the adversarial robustness of the edge computing scenarios is important. Some recently published work can be the reference for the robustness related to the compression and quantization [267, 274, 275].

Thirdly, elucidating the nexus between adversarial robustness and prediction uncertainty emerges as a critical endeavor. Additionally, the PixelDP method [200] offers a pioneering approach that bridges the domains of privacy and robustness. Within the realm of automated driving, the integration of these three pivotal elements, adversarial robustness, prediction uncertainty, and privacy into a cohesive and responsible AI framework, is imperative for comprehensive risk management.

Last but not least, SOTIF evaluation and improvement through the tools of adversarial robustness are another vital way to ensure the safety of automated driving systems. SOTIF underscores the criticality of employing AI technologies and verification strategies adeptly to identify, mitigate, and manage emergent risks. This process necessitates a rigorous analysis of the system's intended functionality, the identification of potential hazard scenarios, and the deployment of measures aimed at mitigating these identified threats. The goal is to ensure that the system is designed and implemented in strict adherence to the safety requirements, with residual risks diminished to an acceptable threshold. For instance, the application of Universal Adversarial Perturbations (UAPs) can critically assess the safety of perception systems within autonomous vehicles. Furthermore, methodologies such as rapid adversarial training techniques [110, 151] can serve as the potential tools within automated driving systems to address immediate risks and enhance model SOTIF. Additionally, certified randomized smoothing methods [20, 213, 217] provide a certified robustness ε-region, offering a foundational model for the development of secure and safe systems within the ICV industry.

8 Conclusion

This survey comprehensively examines the evolution of research on adversarial robustness over the past decade, highlighting the key contributions pertinent to the automated driving systems. Furthermore, it identifies several prospective research achievements, offering preliminary insights aimed at addressing these emergent challenges. Our review synthesizes the existing scholarship with the forward-looking perspectives, positioning itself as a pivotal resource for stakeholders in cybersecurity and SOTIF within the realm of automated driving. Notably, this work does not delve into the theoretical foundations of adversarial vulnerability and robustness in deep learning frameworks, designating this critical area as a subject for future inquiry.

References

1. SAE: Taxonomy and definitions for terms related to driving automation systems for on-road motor vehicles j3016_202104. https://www.sae.org/standards/content/j3016_202104 (2021)
2. Carlini, N.: A complete list of all (arxiv) adversarial example papers. https://nicholas.carlini.com/writing/2019/all-adversarial-example-papers.html (2019)
3. Szegedy, C., Zaremba, W., Sutskever, I., Bruna, J., Erhan, D., Goodfellow, I., Fergus, R.: Intriguing properties of neural networks. In: 2nd International Conference on Learning Representations (ICLR) (2014)
4. Goodfellow, I.J., Shlens, J., Szegedy, C.: Explaining and harnessing adversarial examples. In: 3rd International Conference on Learning Representations (ICLR) (2015)
5. Carlini, N., Wagner, D.: Towards evaluating the robustness of neural networks. In: IEEE Symposium on Security and Privacy (SP), pp. 39–57. IEEE (2017)
6. Madry, A., Makelov, A., Schmidt, L., Tsipras, D., Vladu, A.: Towards deep learning models resistant to adversarial attacks. In: International Conference on Learning Representations (2018)
7. Zhang, H., Yu, Y., Jiao, J., Xing, E., Ghaoui, L.E., Jordan, M.: Theoretically principled trade-off between robustness and accuracy. In: International Conference on Machine Learning (ICML), pp. 7472–7482. PMLR (2019)
8. Dong, Y., Fu, Q.A., Yang, X., Pang, T., Su, H., Xiao, Z., Zhu, J.: Benchmarking adversarial robustness on image classification. In: Proceedings of the IEEE/CVF Conference on Computer Vision and Pattern Recognition (CVPR), pp. 321–331 (2020)
9. Dong, Y., Kang, C., Zhang, J., Zhu, Z., Wang, Y., Yang, X., Su, H., Wei, X., Zhu, J.: Benchmarking robustness of 3D object detection to common corruptions. In: Proceedings of the IEEE/CVF Conference on Computer Vision and Pattern Recognition, pp. 1022–1032 (2023)
10. Tang, S., Gong, R., Wang, Y., Liu, A., Wang, J., Chen, X., Yu, F., Liu, X., Song, D., Yuille, A., et al.: Robustart: benchmarking robustness on architecture design and training techniques (2021). arXiv preprint arXiv:2109.05211
11. Yin, D., Lopes, R.G., Shlens, J., Cubuk, E.D., Gilmer, J.: A Fourier perspective on model robustness in computer vision. Adv. Neural Inf. Process. Syst. 32 (2019)
12. Fawzi, A., Moosavi-Dezfooli, S.M., Frossard, P.: Robustness of classifiers: from adversarial to random noise. Adv. Neural Inf. Process. Syst. 29 (2016)
13. Eykholt, K., Evtimov, I., Fernandes, E., Li, B., Rahmati, A., Xiao, C., Prakash, A., Kohno, T., Song, D.: Robust physical-world attacks on deep learning visual classification. In: Proceedings of the IEEE Conference on Computer Vision and Pattern Recognition, pp. 1625–1634 (2018)

14. Zhang, Y., Foroosh, H., David, P., Gong, B.: Camou: learning physical vehicle camouflages to adversarially attack detectors in the wild. In: International Conference on Learning Representations (2019)
15. Zhang, Q., Hu, S., Sun, J., Chen, Q.A., Mao, Z.M.: On adversarial robustness of trajectory prediction for autonomous vehicles. In: Proceedings of the IEEE/CVF Conference on Computer Vision and Pattern Recognition, pp. 15159–15168 (2022)
16. Cao, Y., Xiao, C., Cyr, B., Zhou, Y., Park, W., Rampazzi, S., Chen, Q.A., Fu, K., Mao, Z.M.: Adversarial sensor attack on lidar-based perception in autonomous driving. In: Proceedings of the 2019 ACM SIGSAC Conference on Computer and Communications Security, pp. 2267–2281 (2019)
17. Sato, T., Shen, J., Wang, N., Jia, Y., Lin, X., Chen, Q.A.: Dirty road can attack: security of deep learning based automated lane centering under Physical-World attack. In: 30th USENIX Security Symposium (USENIX Security 21), pp. 3309–3326 (2021)
18. Ikram, M.H., Khaliq, S., Anjum, M.L., Hussain, W.: Perceptual aliasing++: adversarial attack for visual slam front-end and back-end. IEEE Robot. Autom. Lett. **7**(2), 4670–4677 (2022)
19. Kurakin, A., Goodfellow, I.J., Bengio, S.: Adversarial examples in the physical world. In: Artificial Intelligence Safety and Security, pp. 99–112. Chapman and Hall/CRC (2018)
20. Cohen, J., Rosenfeld, E., Kolter, Z.: Certified adversarial robustness via randomized smoothing. In: International Conference on Machine Learning, pp. 1310–1320. PMLR (2019)
21. Kurakin, A., Goodfellow, I., Bengio, S.: Adversarial machine learning at scale (2016). arXiv preprint arXiv:1611.01236
22. Dong, Y., Liao, F., Pang, T., Su, H., Zhu, J., Hu, X., Li, J.: Boosting adversarial attacks with momentum. In: Proceedings of the IEEE Conference on Computer Vision and Pattern Recognition, pp. 9185–9193 (2018)
23. Papernot, N., McDaniel, P., Jha, S., Fredrikson, M., Celik, Z.B., Swami, A.: The limitations of deep learning in adversarial settings. In: 2016 IEEE European Symposium on Security and Privacy (EuroS&P), pp. 372–387. IEEE (2016)
24. Chen, P.Y., Sharma, Y., Zhang, H., Yi, J., Hsieh, C.J.: Ead: elastic-net attacks to deep neural networks via adversarial examples. In: Proceedings of the AAAI Conference on Artificial Intelligence, vol. 32 (2018)
25. Athalye, A., Engstrom, L., Ilyas, A., Kwok, K.: Synthesizing robust adversarial examples. In: International Conference on Machine Learning, pp. 284–293. PMLR (2018)
26. Athalye, A., Carlini, N., Wagner, D.: Obfuscated gradients give a false sense of security: circumventing defenses to adversarial examples. In: International Conference on Machine Learning, pp. 274–283. PMLR (2018)
27. He, W., Li, B., Song, D.: Decision boundary analysis of adversarial examples. In: International Conference on Learning Representations (2018)
28. Croce, F., Hein, M.: Reliable evaluation of adversarial robustness with an ensemble of diverse parameter-free attacks. In: International Conference on Machine Learning, pp. 2206–2216. PMLR (2020)
29. Moosavi-Dezfooli, S.M., Fawzi, A., Frossard, P.: Deepfool: a simple and accurate method to fool deep neural networks. In: Proceedings of the IEEE Conference on Computer Vision and Pattern Recognition, pp. 2574–2582 (2016)
30. Moosavi-Dezfooli, S.M., Fawzi, A., Fawzi, O., Frossard, P.: Universal adversarial perturbations. In: Proceedings of the IEEE Conference on Computer Vision and Pattern Recognition, pp. 1765–1773 (2017)
31. Hayes, J., Danezis, G.: Learning universal adversarial perturbations with generative models. In: 2018 IEEE Security and Privacy Workshops (SPW), pp. 43–49. IEEE (2018)
32. Baluja, S., Fischer, I.: Learning to attack: adversarial transformation networks. In: Proceedings of the AAAI Conference on Artificial Intelligence, vol. 32 (2018)

33. Mopuri, K.R., Garg, U., Babu, R.V.: Fast feature fool: a data independent approach to universal adversarial perturbations (2017). arXiv preprint arXiv:1707.05572
34. Mopuri, K.R., Ganeshan, A., Babu, R.V.: Generalizable data-free objective for crafting universal adversarial perturbations. IEEE Trans. Pattern Anal. Mach. Intell. **41**(10), 2452–2465 (2018)
35. Hendrycks, D., Dietterich, T.: Benchmarking neural network robustness to common corruptions and perturbations (2019). arXiv preprint arXiv:1903.12261
36. Co, K.T., Muñoz-González, L., de Maupeou, S., Lupu, E.C.: Procedural noise adversarial examples for black-box attacks on deep convolutional networks. In: Proceedings of the 2019 ACM SIGSAC Conference on Computer and Communications Security, pp. 275–289 (2019)
37. Yan, J., Yin, H., Ge, W., Liu, L.: Exploring aesthetic procedural noise for crafting model-agnostic universal adversarial perturbations. Displays **79**, 102479 (2023)
38. Papernot, N., McDaniel, P., Goodfellow, I., Jha, S., Celik, Z.B., Swami, A.: Practical black-box attacks against machine learning. In: Proceedings of the 2017 ACM on Asia Conference on Computer and Communications Security, pp. 506–519 (2017)
39. Shi, Y., Wang, S., Han, Y.: Curls & whey: boosting black-box adversarial attacks. In: Proceedings of the IEEE/CVF Conference on Computer Vision and Pattern Recognition, pp. 6519–6527 (2019)
40. Dong, Y., Pang, T., Su, H., Zhu, J.: Evading defenses to transferable adversarial examples by translation-invariant attacks. In: Proceedings of the IEEE/CVF Conference on Computer Vision and Pattern Recognition, pp. 4312–4321 (2019)
41. Xie, C., Zhang, Z., Zhou, Y., Bai, S., Wang, J., Ren, Z., Yuille, A.L.: Improving transferability of adversarial examples with input diversity. In: Proceedings of the IEEE/CVF Conference on Computer Vision and Pattern Recognition, pp. 2730–2739 (2019)
42. Wang, X., He, K.: Enhancing the transferability of adversarial attacks through variance tuning. In: Proceedings of the IEEE/CVF Conference on Computer Vision and Pattern Recognition, pp. 1924–1933 (2021)
43. Huang, Z., Zhang, T.: Black-box adversarial attack with transferable model-based embedding (2019). arXiv preprint arXiv:1911.07140
44. Qin, Z., Fan, Y., Liu, Y., Shen, L., Zhang, Y., Wang, J., Baoyuan, W.: Boosting the transferability of adversarial attacks with reverse adversarial perturbation. Adv. Neural. Inf. Process. Syst. **35**, 29845–29858 (2022)
45. Yang, X., Dong, Y., Pang, T., Su, H., Zhu, J.: Boosting transferability of targeted adversarial examples via hierarchical generative networks. In: European Conference on Computer Vision, pp. 725–742. Springer (2022)
46. Ilyas, A., Engstrom, L., Athalye, A., Lin, J.: Black-box adversarial attacks with limited queries and information. In: International Conference on Machine Learning, pp. 2137–2146. PMLR (2018)
47. Li, Y., Li, L., Wang, L., Zhang, T., Gong, B.: Nattack: learning the distributions of adversarial examples for an improved black-box attack on deep neural networks. In: International Conference on Machine Learning, pp. 3866–3876. PMLR (2019)
48. Mohaghegh Dolatabadi, H., Erfani, S., Leckie, S.: Advflow: inconspicuous black-box adversarial attacks using normalizing flows. Adv. Neural Inf. Process. Syst. **33**, 15871–15884 (2020)
49. Liu, S., Chen, P.Y., Chen, X., Hong, M.: signsgd via zeroth-order oracle. In: International Conference on Learning Representations ICLR (2019)
50. Ilyas, A., Engstrom, L., Madry, A.: Prior convictions: black-box adversarial attacks with bandits and priors (2018). arXiv preprint arXiv:1807.07978
51. Guo, C., Gardner, J., You, Y., Wilson, A.G., Weinberger, K.: Simple black-box adversarial attacks. In: International Conference on Machine Learning, pp. 2484–2493. PMLR (2019)

52. Moon, S., An, G., Song, H.O.: Parsimonious black-box adversarial attacks via efficient combinatorial optimization. In: International Conference on Machine Learning, pp. 4636–4645. PMLR (2019)
53. Al-Dujaili, A., O'Reilly, U.M.: Sign bits are all you need for black-box attacks. In: International Conference on Learning Representations (2020)
54. Andriushchenko, M., Croce, F., Flammarion, N., Hein, M.: Square attack: a query-efficient black-box adversarial attack via random search. In: European Conference on Computer Vision, pp. 484–501. Springer (2020)
55. Feng, Y., Wu, B., Fan, Y., Liu, L., Li, Z., Xia, S.T.: Boosting black-box attack with partially transferred conditional adversarial distribution. In: Proceedings of the IEEE/CVF Conference on Computer Vision and Pattern Recognition, pp. 15095–15104 (2022)
56. Brendel, W., Rauber, J., Bethge, M.: Decision-based adversarial attacks: reliable attacks against black-box machine learning models. In: International Conference on Learning Representations (2018)
57. Cheng, M., Le, T., Chen, P.Y., Zhang, H., Yi, J.F., Hsieh, C.J.: Query-efficient hard-label black-box attack: an optimization-based approach. In: International Conference on Learning Representations (2019)
58. Cheng, M., Singh, S., Chen, P.H., Chen, P.Y., Liu, S., Hsieh, C.J.: Sign-opt: a query-efficient hard-label adversarial attack. In: International Conference on Learning Representations (2020)
59. Dong, Y., Su, H., Wu, B., Li, Z., Liu, W., Zhang, T., Zhu, J.: Efficient decision-based black-box adversarial attacks on face recognition. In: Proceedings of the IEEE/CVF Conference on Computer Vision and Pattern Recognition, pp. 7714–7722 (2019)
60. Shi, Y., Han, Y., Qinghua, H., Yang, Y., Tian, Q.: Query-efficient black-box adversarial attack with customized iteration and sampling. IEEE Trans. Pattern Anal. Mach. Intell. **45**(2), 2226–2245 (2022)
61. Rahmati, A., Moosavi-Dezfooli, S.M., Frossard, P., Dai, H.: Geoda: a geometric framework for black-box adversarial attacks. In: Proceedings of the IEEE/CVF Conference on Computer Vision and Pattern Recognition, pp. 8446–8455 (2020)
62. Chen, J., Jordan, M.I., Wainwright, M.J.: Hopskipjumpattack: a query-efficient decision-based attack. In: 2020 IEEE Symposium on Security and Privacy (sp), pp. 1277–1294. IEEE (2020)
63. Li, H., Xu, X., Zhang, X., Yang, S., Li, B.: Qeba: query-efficient boundary-based blackbox attack. In: Proceedings of the IEEE/CVF Conference on Computer Vision and Pattern Recognition, pp. 1221–1230 (2020)
64. Chen, W., Zhang, Z., Hu, X., Wu, B.: Boosting decision-based black-box adversarial attacks with random sign flip. In: European Conference on Computer Vision, pp. 276–293. Springer (2020)
65. Chen, J., Gu, Q.: Rays: a ray searching method for hard-label adversarial attack. In: Proceedings of the 26th ACM SIGKDD International Conference on Knowledge Discovery & Data Mining, pp. 1739–1747 (2020)
66. Liu, S., Chen, P.Y., Kailkhura, B., Zhang, G., Hero III, A.O., Varshney, P.V.: A primer on zeroth-order optimization in signal processing and machine learning: principals, recent advances, and applications. IEEE Signal Process. Mag. **37**(5), 43–54 (2020)
67. Liang, S., Wu, B., Fan, Y., Wei, X., Cao, X.: Parallel rectangle flip attack: a query-based black-box attack against object detection. In: International Conference on Computer Vision (2021)
68. Yan, J., Yin, H., Zhao, Z., Ge, W., Zhang, H., Rigoll, G.: Wavelet regularization benefits adversarial training. Inf. Sci. **649**, 119650 (2023)
69. Zheng, M., Yan, X., Zhu, Z., Chen, H., Wu, B.: Blackboxbench: a comprehensive benchmark of black-box adversarial attacks (2023). arXiv preprint arXiv:2312.16979
70. Zagoruyko, S., Komodakis, N.: Wide residual networks. In: British Conference on Computer Vision (2016)

71. Dosovitskiy, A., Beyer, L., Kolesnikov, A., Weissenborn, D., Zhai, X., Unterthiner, T., Dehghani, M., Minderer, M., Heigold, G., Gelly, S., et al.: An image is worth 16x16 words: transformers for image recognition at scale. In: International Conference on Learning Representations (2021)
72. Liu, Z., Lin, Y., Cao, Y., Hu, H., Wei, Y., Zhang, Z., Lin, S., Guo, B.: Swin transformer: hierarchical vision transformer using shifted windows. In: Proceedings of the IEEE/CVF International Conference on Computer Vision, pp. 10012–10022 (2021)
73. Song, D., Eykholt, K., Evtimov, I., Fernandes, E., Li, B., Rahmati, A., Tramer, F., Prakash, A., Kohno, T.: Physical adversarial examples for object detectors. In: 12th USENIX Workshop on Offensive Technologies (WOOT 18) (2018)
74. Li, J., Schmidt, F., Kolter, Z.: Adversarial camera stickers: a physical camera-based attack on deep learning systems. In: International Conference on Machine Learning, pp. 3896–3904. PMLR (2019)
75. Zhao, Y., Zhu, H., Liang, R., Shen, Q., Zhang, S., Chen, K.: Seeing isn't believing: towards more robust adversarial attack against real world object detectors. In: Proceedings of the 2019 ACM SIGSAC Conference on Computer and Communications Security, pp. 1989–2004 (2019)
76. Duan, R., Ma, X., Wang, Y., Bailey, J., Qin, A.K., Yang, Y.: Adversarial camouflage: hiding physical-world attacks with natural styles. In: Proceedings of the IEEE/CVF Conference on Computer Vision and Pattern Recognition, pp. 1000–1008 (2020)
77. Kong, Z., Guo, J., Li, A., Liu, C.: Physgan: generating physical-world-resilient adversarial examples for autonomous driving. In: Proceedings of the IEEE/CVF Conference on Computer Vision and Pattern Recognition, pp. 14254–14263 (2020)
78. Wu, T., Ning, X., Li, W., Huang, R., Yang, H., Wang, Y.: Physical adversarial attack on vehicle detector in the carla simulator (2020). arXiv preprint arXiv:2007.16118
79. Huang, L., Gao, C., Zhou, Y., Xie, C., Yuille, A.L., Zou, C., Liu, N.: Universal physical camouflage attacks on object detectors. In: Proceedings of the IEEE/CVF Conference on Computer Vision and Pattern Recognition, pp. 720–729 (2020)
80. Wu, Z., Lim, S.N., Davis, L.S., Goldstein, T.: Making an invisibility cloak: real world adversarial attacks on object detectors. In: Computer Vision-ECCV 2020: 16th European Conference, Glasgow, UK, 23–28 Aug 2020, Proceedings, Part IV 16, pp. 1–17. Springer (2020)
81. Xu, K., Zhang, G., Liu, S., Fan, Q., Sun, M., Chen, H., Chen, P.Y., Wang, Y., Lin, X.: Adversarial t-shirt! evading person detectors in a physical world. In: Computer Vision-ECCV 2020: 16th European Conference, Glasgow, UK, 23–28 Aug 2020, Proceedings, Part V 16, pp. 665–681. Springer (2020)
82. Boloor, A., Garimella, K., He, X., Gill, C., Vorobeychik, Y., Zhang, X.: Attacking vision-based perception in end-to-end autonomous driving models. J. Syst. Architect. **110**, 101766 (2020)
83. Yamanaka, K., Matsumoto, R., Takahashi, K., Fujii, T.: Adversarial patch attacks on monocular depth estimation networks. IEEE Access **8**, 179094–179104 (2020)
84. Sun, J., Cao, Y., Chen, Q.A., Mao, Z.M.: Towards robust LiDAR-based perception in autonomous driving: General black-box adversarial sensor attack and countermeasures. In: 29th USENIX Security Symposium (USENIX Security 20), pp. 877–894 (2020)
85. Tu, J., Ren, M., Manivasagam, S., Liang, M., Yang, B., Du, R., Cheng, F., Urtasun, R.: Physically realizable adversarial examples for lidar object detection. In: Proceedings of the IEEE/CVF Conference on Computer Vision and Pattern Recognition, pp. 13716–13725 (2020)
86. Bai, T., Luo, J., Zhao, J.: Inconspicuous adversarial patches for fooling image-recognition systems on mobile devices. IEEE Internet Things J. **9**(12), 9515–9524 (2021)
87. Ye, B., Yin, H., Yan, J., Ge, W.: Patch-based attack on traffic sign recognition. In: 2021 IEEE International Intelligent Transportation Systems Conference (ITSC), pp. 164–171. IEEE (2021)
88. Lovisotto, G., Turner, H., Sluganovic, I., Strohmeier, M., Martinovic, I.: SLAP: improving physical adversarial examples with short-lived adversarial perturbations. In: 30th USENIX Security Symposium (USENIX Security 21), pp. 1865–1882 (2021)

89. Duan, R., Mao, X., Qin, A.K., Chen, Y., Ye, S., He, Y., Yang, Y.: Adversarial laser beam: effective physical-world attack to dnns in a blink. In: Proceedings of the IEEE/CVF Conference on Computer Vision and Pattern Recognition, pp. 16062–16071 (2021)
90. Sayles, A., Hooda, A., Gupta, M., Chatterjee, R., Fernandes, E.: Invisible perturbations: physical adversarial examples exploiting the rolling shutter effect. In: Proceedings of the IEEE/CVF Conference on Computer Vision and Pattern Recognition, pp. 14666–14675 (2021)
91. Wang, J., Liu, A., Yin, Z., Liu, S., Tang, S., Liu, X.: Dual attention suppression attack: Generate adversarial camouflage in physical world. In: Proceedings of the IEEE/CVF Conference on Computer Vision and Pattern Recognition, pp. 8565–8574 (2021)
92. Zolfi, A., Kravchik, M., Elovici, Y., Shabtai, A.: The translucent patch: a physical and universal attack on object detectors. In: Proceedings of the IEEE/CVF Conference on Computer Vision and Pattern Recognition, pp. 15232–15241 (2021)
93. Cao, Y., Wang, N., Xiao, C., Yang, D., Fang, J., Yang, R., Chen, Q.A., Liu, M., Li, B.: Invisible for both camera and lidar: Security of multi-sensor fusion based perception in autonomous driving under physical-world attacks. In: 2021 IEEE Symposium on Security and Privacy (SP), pp. 176–194. IEEE (2021)
94. Zhong, Y., Liu, X., Zhai, D., Jiang, J., Ji, X.: Shadows can be dangerous: stealthy and effective physical-world adversarial attack by natural phenomenon. In: Proceedings of the IEEE/CVF Conference on Computer Vision and Pattern Recognition, pp. 15345–15354 (2022)
95. Suryanto, N., Kim, Y., Kang, H., Larasati, H.T., Yun, Y., Le, T.T.H., Yang, H., Oh, S.Y., Kim, H.: Dta: physical camouflage attacks using differentiable transformation network. In: Proceedings of the IEEE/CVF Conference on Computer Vision and Pattern Recognition, pp. 15305–15314 (2022)
96. Cheng, Z., Liang, J., Choi, H., Tao, G., Cao, Z., Liu, D., Zhang, X.: Physical attack on monocular depth estimation with optimal adversarial patches. In: European Conference on Computer Vision, pp. 514–532. Springer (2022)
97. Wan, Z., Shen, J., Chuang, J., Xia, X., Garcia, J., Ma, J., Chen, Q.A.: Too afraid to drive: Systematic discovery of semantic dos vulnerability in autonomous driving planning under physical-world attacks. In: ISOC Network and Distributed Systems Security (NDSS) Symposium (2022)
98. Yan, J., Yin, H., Ye, B., Ge, W., Zhang, H., Rigoll, G.: An adversarial attack on salient regions of traffic sign. Automot. Innov., 1–14 (2023)
99. Cao, Y., Bhupathiraju, S.H., Naghavi, P., Sugawara, T., Mao, Z.M., Rampazzi, S.: You can't see me: physical removal attacks on LiDAR-based autonomous vehicles driving frameworks. In: 32nd USENIX Security Symposium (USENIX Security 23), pp. 2993–3010 (2023)
100. Sharif, M., Bhagavatula, S., Bauer, L., Reiter, M.K.: Accessorize to a crime: real and stealthy attacks on state-of-the-art face recognition. In: Proceedings of the 2016 ACM SIGSAC Conference on Computer and Communications Security, pp. 1528–1540 (2016)
101. Goodfellow, I., Pouget-Abadie, J., Mirza, M., Xu, B., Warde-Farley, D., Ozair, S., Courville, A., Bengio, Y.: Generative adversarial nets. Adv. Neural Inf. Process. Syst. **27** (2014)
102. Brown, T.B., Mané, D., Roy, A., Abadi, M., Gilmer, J.: Adversarial patch (2017). arXiv preprint arXiv:1712.09665
103. Zhou, B., Khosla, A., Lapedriza, A., Oliva, A., Torralba, A.: Learning deep features for discriminative localization. In: Proceedings of the IEEE Conference on Computer Vision and Pattern Recognition, pp. 2921–2929 (2016)
104. Szegedy, C., Liu, W., Jia, Y., Sermanet, P., Reed, S., Anguelov, D., Erhan, D., Vanhoucke, V., Rabinovich, A.: Going deeper with convolutions. In: Proceedings of the IEEE Conference on Computer Vision and Pattern Recognition, pp. 1–9 (2015)
105. He, K., Zhang, X., Ren, S., Sun, J.: Deep residual learning for image recognition. In: Proceedings of the IEEE Conference on Computer Vision and Pattern Recognition, pp. 770–778 (2016)

106. He, K., Gkioxari, G., Dollár, P., Girshick, R.: Mask r-cnn. In: Proceedings of the IEEE International Conference on Computer Vision, pp. 2961–2969 (2017)
107. Redmon, J., Farhadi, A.: Yolov3: an incremental improvement (2018). arXiv preprint arXiv:1804.02767
108. Lang, A.H., Vora, S., Caesar, H., Zhou, L., Yang, J., Beijbom, O.: Pointpillars: fast encoders for object detection from point clouds. In: Proceedings of the IEEE/CVF Conference on Computer Vision and Pattern Recognition, pp. 12697–12705 (2019)
109. Shi, S., Wang, X., Li, H.: Pointrcnn: 3d object proposal generation and detection from point cloud. In: Proceedings of the IEEE/CVF Conference on Computer Vision and Pattern Recognition, pp. 770–779 (2019)
110. Wong, E., Rice, L., Kolter, J.Z.: Fast is better than free: revisiting adversarial training. In: International Conference on Learning Representations (2020)
111. Kuang, X., Zhao, F., Hao, H., Liu, Z.: Intelligent connected vehicles: the industrial practices and impacts on automotive value-chains in china. Asia Pac. Bus. Rev. **24**(1), 1–21 (2018)
112. Geiger, A., Lenz, P., Urtasun, R.: Are we ready for autonomous driving? the kitti vision benchmark suite. In: 2012 IEEE Conference on Computer Vision and Pattern Recognition, pp. 3354–3361. IEEE (2012)
113. Papernot, N., McDaniel, P., Wu, X., Jha, S., Swami, A.: Distillation as a defense to adversarial perturbations against deep neural networks. In: 2016 IEEE Symposium on Security and Privacy (SP), pp. 582–597. IEEE (2016)
114. Tramèr, F., Kurakin, A., Papernot, N., Goodfellow, I., Boneh, D., McDaniel, P.: Ensemble adversarial training: attacks and defenses. In: International Conference on Learning Representations (2018)
115. Yan, Z., Guo, Y., Zhang, C.: Deep defense: training dnns with improved adversarial robustness. Adv. Neural Inf. Process. Syst., 31 (2018)
116. Song, C., He, K., Wang, L., Hopcroft, J.E.: Improving the generalization of adversarial training with domain adaptation. In: International Conference on Learning Representations (2019)
117. Wang, Y., Zou, D., Yi, J., Bailey, J., Ma, X., Gu, Q.: Improving adversarial robustness requires revisiting misclassified examples. In: International Conference on Learning Representations (2020)
118. Stutz, D., Hein, M., Schiele, B.: Confidence-calibrated adversarial training: generalizing to unseen attacks. In: International Conference on Machine Learning, pp. 9155–9166. PMLR (2020)
119. Zhang, J., Xu, X., Han, B., Niu, G., Cui, L., Sugiyama, M., Kankanhalli, M.: Attacks which do not kill training make adversarial learning stronger. In: International Conference on Machine Learning, pp. 11278–11287. PMLR (2020)
120. Yang, H., Zhang, J., Dong, H., Inkawhich, N., Gardner, A., Touchet, A., Wilkes, W., Berry, H., Li, H.: Dverge: diversifying vulnerabilities for enhanced robust generation of ensembles. Adv. Neural Inf. Process. Syst. **33**, 5505–5515 (2020)
121. Pang, T., Yang, X., Dong, Y., Su, H., Zhu, J.: Bag of tricks for adversarial training. In: International Conference on Learning Representations (2021)
122. Wang, Z., Pang, T., Du, C., Lin, M., Liu, W., Yan, S.: Better diffusion models further improve adversarial training. In: International Conference on Machine Learning (2023)
123. Yan, J., Yin, H., Zhao, Z., Ge, W., Zhang, J.: Enhance adversarial robustness via geodesic distance. IEEE Trans. Artif. Intell. **5**(8), 4202–4216 (2024)
124. Metzen, J.H., Genewein, T., Fischer, V., Bischoff, B.: On detecting adversarial perturbations. In: International Conference on Learning Representations (2016)
125. Lu, J., Issaranon, T., Forsyth, D.: Safetynet: detecting and rejecting adversarial examples robustly. In: Proceedings of the IEEE International Conference on Computer Vision, pp. 446–454 (2017)

126. Meng, D., Chen, H.: Magnet: a two-pronged defense against adversarial examples. In: Proceedings of the 2017 ACM SIGSAC Conference on Computer and Communications Security, pp. 135–147 (2017)
127. Zheng, Z., Hong, P.: Robust detection of adversarial attacks by modeling the intrinsic properties of deep neural networks. Adv. Neural Inf. Process. Syst. **31** (2018)
128. Lee, K., Lee, K., Lee, H., Shin, J.: A simple unified framework for detecting out-of-distribution samples and adversarial attacks. Adv. Neural Inf. Process. Syst. **31** (2018)
129. Pang, T., Du, C., Dong, Y., Zhu, J.: Towards robust detection of adversarial examples. Adv. Neural Inf. Process. Syst. **31** (2018)
130. Yang, K., Zhou, T., Zhang, Y., Tian, X., Tao, D.: Class-disentanglement and applications in adversarial detection and defense. Adv. Neural Inf. Process. Syst. **34**, 16051–16063 (2021)
131. Deng, Z., Yang, X., Xu, S., Su, H., Zhu, J.: Libre: a practical Bayesian approach to adversarial detection. In: Proceedings of the IEEE/CVF Conference on Computer Vision and Pattern Recognition, pp. 972–982 (2021)
132. Li, H., Shan, S., Wenger, E., Zhang, J., Zheng, H., Zhao, B.Y.: Blacklight: scalable defense for neural networks against Query-BasedBlack-Box attacks. In: 31st USENIX Security Symposium (USENIX Security 22), pp. 2117–2134 (2022)
133. Juuti, M., Szyller, S., Marchal, S., Asokan, N.: Prada: protecting against dnn model stealing attacks. In: 2019 IEEE European Symposium on Security and Privacy (EuroS&P), pp. 512–527. IEEE (2019)
134. Chen, S., Carlini, N., Wagner, D.: Stateful detection of black-box adversarial attacks. In: Proceedings of the 1st ACM Workshop on Security and Privacy on Artificial Intelligence, pp. 30–39 (2020)
135. Guo, C., Rana, M., Cisse, M., van der Maaten, L.: Countering adversarial images using input transformations. In: International Conference on Learning Representations (2018)
136. Song, Y., Kim, T., Nowozin, S., Ermon, S., Kushman, N.: Pixeldefend: leveraging generative models to understand and defend against adversarial examples. In: International Conference on Learning Representations (2018)
137. Xie, C., Wang, J., Zhang, Z., Ren, Z., Yuille, A.: Mitigating adversarial effects through randomization. In: International Conference on Learning Representations (2018)
138. Raff, E., Sylvester, J., Forsyth, S., McLean, M.: Barrage of random transforms for adversarially robust defense. In: Proceedings of the IEEE/CVF Conference on Computer Vision and Pattern Recognition, pp. 6528–6537 (2019)
139. Pang, T., Xu, K., Zhu, J.: Mixup inference: better exploiting mixup to defend adversarial attacks. In: International Conference on Learning Representations (2020)
140. Qin, Z., Fan, Y., Zha, H., Baoyuan, W.: Random noise defense against query-based black-box attacks. Adv. Neural Inf. Process. Syst. **34**, 7650–7663 (2021)
141. Nie, W., Guo, B., Huang, Y., Xiao, C., Vahdat, A., Anandkumar, A.: Diffusion models for adversarial purification. In: International Conference on Machine Learning, pp. 16805–16827. PMLR (2022)
142. Chen, S., Huang, Z., Tao, Q., Yingwen, W., Xie, C., Huang, X.: Adversarial attack on attackers: Post-process to mitigate black-box score-based query attacks. Adv. Neural Inf. Process. Syst. **35**, 14929–14943 (2022)
143. Aithal, M.B., Li, X.: Boundary defense against black-box adversarial attacks. In: 2022 26th International Conference on Pattern Recognition (ICPR), pp. 2349–2356. IEEE (2022)
144. Alfarra, M., Pérez, J.C., Thabet, A., Bibi, A., Torr, P.H.S., Ghanem, B.: Combating adversaries with anti-adversaries. In: Proceedings of the AAAI Conference on Artificial Intelligence, volume 36, pp. 5992–6000 (2022)
145. Wang, D., Ju, A., Shelhamer, E., Wagner, D., Darrell, T.: Fighting gradients with gradients: dynamic defenses against adversarial attacks (2021). arXiv preprint arXiv:2105.08714

146. Yoon, J., Hwang, S.J., Lee, J.: Adversarial purification with score-based generative models. In: International Conference on Machine Learning, pp. 12062–12072. PMLR (2021)
147. Shi, C., Holtz, C., Mishne, G.: Online adversarial purification based on self-supervised learning. In: International Conference on Learning Representations (2021)
148. Shafahi, A., Najibi, M., Ghiasi, M.A., Xu, Z., Dickerson, J., Studer, C., Davis, L.S., Taylor, G., Goldstein, T.: Adversarial training for free! Adv. Neural Inf. Process. Syst. **32** (2019)
149. Zhang, D., Zhang, T., Lu, Y., Zhu, Z., Dong, B.: You only propagate once: accelerating adversarial training via maximal principle. Adv. Neural Inf. Process. Syst. **32** (2019)
150. Qin, C., Martens, J., Gowal, S., Krishnan, D., Dvijotham, K., Fawzi, A., De, S., Stanforth, R., Kohli, P.: Adversarial robustness through local linearization. Adv. Neural Inf. Process. Syst. **32** (2019)
151. Andriushchenko, M., Flammarion, N.: Understanding and improving fast adversarial training. Adv. Neural Inf. Process. Syst. **33**, 16048–16059 (2020)
152. Grabinski, J., Jung, S., Keuper, J., Keuper, M.: Frequency lowcut pooling-plug and play against catastrophic overfitting. In: European Conference on Computer Vision, pp. 36–57. Springer (2022)
153. Zhu, K., Hu, X., Wang, J., Xie, X., Yang, G.: Improving generalization of adversarial training via robust critical fine-tuning. In: Proceedings of the IEEE/CVF International Conference on Computer Vision, pp. 4424–4434 (2023)
154. Xie, C., Wu, Y., van der Maaten, L., Yuille, A.L., He, K.: Feature denoising for improving adversarial robustness. In: Proceedings of the IEEE/CVF Conference on Computer Vision and Pattern Recognition, pp. 501–509 (2019)
155. Krizhevsky, A., et al.: Learning multiple layers of features from tiny images (2009)
156. Deng, J., Dong, W., Socher, R., Li, L.J., Li, K., Fei-Fei, L.: Imagenet: a large-scale hierarchical image database. In: 2009 IEEE Conference on Computer Vision and Pattern Recognition, pp. 248–255. IEEE (2009)
157. Pang, T., Xu, K., Du C., Chen, N., Zhu, J.: Improving adversarial robustness via promoting ensemble diversity. In: International Conference on Machine Learning, pp. 4970–4979. PMLR (2019)
158. Buades, A., Coll, B., Morel, J.M.: A non-local algorithm for image denoising. In: 2005 IEEE Computer Society Conference on Computer Vision and Pattern Recognition (CVPR'05), vol. 2, pp. 60–65. IEEE (2005)
159. Yang, Y., Zhang, G., Katabi, D., Xu, Z.: Me-net: towards effective adversarial robustness with matrix estimation. In: International Conference on Machine Learning, pp. 7025–7034. PMLR (2019)
160. Alayrac, J.B., Uesato, J., Huang, P.S., Fawzi, A., Stanforth, R., Kohli, P.: Are labels required for improving adversarial robustness? Adv. Neural Inf. Process. Syst. **32** (2019)
161. Balunovic, M., Vechev, M.: Adversarial training and provable defenses: bridging the gap. In: International Conference on Learning Representations (2020)
162. Chan, A., Tay, Y., Ong, Y.S., Fu, J.: Jacobian adversarially regularized networks for robustness. In: International Conference on Learning Representations (2020)
163. Chan, A., Ong, Y.S., Tan, C.: How does frequency bias affect the robustness of neural image classifiers against common corruption and adversarial perturbations? (2022). arXiv preprint arXiv:2205.04533
164. Chen, Z., Li, Q., Zhang, Z.: Towards robust neural networks via close-loop control. In: International Conference on Learning Representations (2021)
165. Tsipras, D., Santurkar, S., Engstrom, L., Turner, A., Madry, A.: Robustness may be at odds with accuracy. In: International Conference on Learning Representations (2019)

166. Zhang, J., Zhu, J., Niu, G., Han, B., Sugiyama, M., Kankanhalli, M.: Geometry-aware instance-reweighted adversarial training. In: International Conference on Learning Representations (2021)
167. Dong, Y., Deng, Z., Pang, T., Zhu, J., Hang, S.: Adversarial distributional training for robust deep learning. Adv. Neural Inf. Process. Syst. **33**, 8270–8283 (2020)
168. Dongxian, W., Xia, S.-T., Wang, Y.: Adversarial weight perturbation helps robust generalization. Adv. Neural Inf. Process. Syst. **33**, 2958–2969 (2020)
169. Rebuffi, S.A., Gowal, S., Calian, D.A., Stimberg, F., Wiles, O., Mann, T.A.: Data augmentation can improve robustness. Adv. Neural Inf. Process. Syst. **34**, 29935–29948 (2021)
170. Pang, T., Lin, M., Yang, X., Zhu, J., Yan, S.: Robustness and accuracy could be reconcilable by (proper) definition. In: International Conference on Machine Learning, pp. 17258–17277. PMLR (2022)
171. Jia, X., Zhang, Y., Wu, B., Ma, K., Wang, J., Cao, X.: Las-at: adversarial training with learnable attack strategy. In: Proceedings of the IEEE/CVF Conference on Computer Vision and Pattern Recognition, pp. 13398–13408 (2022)
172. Williams, R.J.: Simple statistical gradient-following algorithms for connectionist reinforcement learning. Mach. Learn. **8**, 229–256 (1992)
173. Cordts, M., Omran, M., Ramos, S., Rehfeld, T., Enzweiler, M., Benenson, R., Franke, U., Roth, S., Schiele, B.: The cityscapes dataset for semantic urban scene understanding. In: Proceedings of the IEEE Conference on Computer Vision and Pattern Recognition, pp. 3213–3223 (2016)
174. Kang, Q., Song, Y., Ding, Q., Tay, W.P.: Stable neural ode with lyapunov-stable equilibrium points for defending against adversarial attacks. Adv. Neural Inf. Process. Syst. **34**, 14925–14937 (2021)
175. Sautier, C., Puy, G., Gidaris, S., Boulch, A., Bursuc, A., Marlet, R.: Image-to-lidar self-supervised distillation for autonomous driving data. In Proceedings of the IEEE/CVF Conference on Computer Vision and Pattern Recognition, pp. 9891–9901 (2022)
176. Zhou, S., Liu, W., Hu, C., Zhou, S., Ma, C.: Unidistill: a universal cross-modality knowledge distillation framework for 3d object detection in bird's-eye view. In Proceedings of the IEEE/CVF Conference on Computer Vision and Pattern Recognition, pp. 5116–5125 (2023)
177. Song, Y., Sohl-Dickstein, J., Kingma, D.P., Kumar, A., Ermon, S., Poole, B.: Score-based generative modeling through stochastic differential equations. In: International Conference on Learning Representations (2021)
178. Ma, X., Li, B., Wang, Y., Erfani, S.M., Wijewickrema, S., Schoenebeck, G., Song, D., Houle, M.E., Bailey, J.: Characterizing adversarial subspaces using local intrinsic dimensionality. In: International Conference on Learning Representations (2018)
179. Katz, G., Barrett, C., Dill, D.L., Julian, K., Kochenderfer, MJ.: Reluplex: an efficient SMT solver for verifying deep neural networks. In: Computer Aided Verification: 29th International Conference, CAV 2017, Heidelberg, Germany, 24–28 July 2017, Proceedings, Part I 30, pp. 97–117. Springer (2017)
180. Huang, X., Kwiatkowska, M., Wang, S., Wu, M.: Safety verification of deep neural networks. In: Computer Aided Verification: 29th International Conference, CAV 2017, Heidelberg, Germany, 24–28 July 2017, Proceedings, Part I 30, pp. 3–29. Springer (2017)
181. Ehlers, R.: Formal verification of piece-wise linear feed-forward neural networks. In: Automated Technology for Verification and Analysis: 15th International Symposium, ATVA 2017, Pune, India, 3–6 Oct 2017, Proceedings 15, pp. 269–286. Springer (2017)
182. Cheng, C.H., Nührenberg, G., Ruess, H.: Maximum resilience of artificial neural networks. In: Automated Technology for Verification and Analysis: 15th International Symposium, ATVA 2017, Pune, India, 3–6 Oct 2017, Proceedings 15, pp. 251–268. Springer (2017)
183. Hein, M., Andriushchenko, M.: Formal guarantees on the robustness of a classifier against adversarial manipulation. Adv. Neural Inf. Process. Syst. **30** (2017)

184. Xiang, W., Tran, H.D., Johnson, T.T.: Output reachable set estimation and verification for multilayer neural networks. IEEE Trans. Neural Netw. Learn. Syst. **29**(11), 5777–5783 (2018)
185. Bunel, R.R., Turkaslan, I., Torr, P., Kohli, P., Mudigonda, P.K.: A unified view of piecewise linear neural network verification. Adv. Neural Inf. Process. Syst. **31** (2018)
186. Wong, E., Kolter, Z.: Provable defenses against adversarial examples via the convex outer adversarial polytope. In: International Conference on Machine Learning, pp. 5286–5295. PMLR (2018)
187. Dvijotham, K., Stanforth, R., Gowal, S., Mann, T.A., Kohli, P.: A dual approach to scalable verification of deep networks. In: UAI, vol. 1, p. 3 (2018)
188. Raghunathan, A., Steinhardt, J., Liang, P.S.: Semidefinite relaxations for certifying robustness to adversarial examples. Adv. Neural Inf. Process. Syst. **31** (2018)
189. Wang, S., Pei, K., Whitehouse, J., Yang, J., Jana, S.: Formal security analysis of neural networks using symbolic intervals. In: 27th USENIX Security Symposium (USENIX Security 18), pp. 1599–1614 (2018)
190. Wang, S., Pei, K., Whitehouse, J., Yang, J., Jana, S.: Efficient formal safety analysis of neural networks. Adv. Neural Inf. Process. Syst. **31** (2018)
191. Weng, L., Zhang, H., Chen, H., Song, Z., Hsieh, C.J., Daniel, L., Boning, D., Dhillon, I.: Towards fast computation of certified robustness for relu networks. In: International Conference on Machine Learning, pp. 5276–5285. PMLR (2018)
192. Tsuzuku, Y., Sato, I., Sugiyama, M.: Lipschitz-margin training: scalable certification of perturbation invariance for deep neural networks. Adv. Neural Inf. Process. Syst. **31** (2018)
193. Richards, S.M., Berkenkamp, F., Krause, A.: The lyapunov neural network: adaptive stability certification for safe learning of dynamical systems. In: Conference on Robot Learning, pp. 466–476. PMLR (2018)
194. Fazlyab, M., Robey, A., Hassani, H., Morari, M., Pappas, G.: Efficient and accurate estimation of lipschitz constants for deep neural networks. Adv. Neural Inf. Process. Syst. **32** (2019)
195. Jordan, M., Lewis, J., Dimakis, A.G.: Provable certificates for adversarial examples: Fitting a ball in the union of polytopes. Adv. Neural Inf. Process. Syst. **32** (2019)
196. Salman, H., Yang, G., Zhang, H., Hsieh, C.J., Zhang, P.: A convex relaxation barrier to tight robustness verification of neural networks. Adv. Neural Inf. Process. Syst. **32** (2019)
197. Gowal, S., Dvijotham, K.D., Stanforth, R., Bunel, R., Qin, C., Uesato, J., Arandjelovic, R., Mann, T., Kohli, P.: Scalable verified training for provably robust image classification. In: Proceedings of the IEEE/CVF International Conference on Computer Vision, pp. 4842–4851 (2019)
198. Croce, F., Andriushchenko, M., Hein, M.: Provable robustness of relu networks via maximization of linear regions. In: 22nd International Conference on Artificial Intelligence and Statistics, pp. 2057–2066. PMLR (2019)
199. Croce, F., Hein, M.: Provable robustness against all adversarial l_p-perturbations for $p \geq 1$. In: International Conference on Learning Representations (2020)
200. Lecuyer, M., Atlidakis, V., Geambasu, R., Hsu, D., Jana, S.: Certified robustness to adversarial examples with differential privacy. In: 2019 IEEE Symposium on Security and Privacy (SP), pp. 656–672. IEEE (2019)
201. Salman, H., Li, J., Razenshteyn, I., Zhang, P., Zhang, H., Bubeck, S., Yang, G.: Provably robust deep learning via adversarially trained smoothed classifiers. Adv. Neural Inf. Process. Syst. **32** (2019)
202. Mangal, R., Nori, A.V., Orso, A.: Robustness of neural networks: a probabilistic and practical approach. In: 2019 IEEE/ACM 41st International Conference on Software Engineering: New Ideas and Emerging Results (ICSE-NIER), pp. 93–96. IEEE (2019)

203. Weng, L., Chen, P.Y., Nguyen, L., Squillante, M., Boopathy, A., Oseledets, I., Daniel, L.: Proven: verifying robustness of neural networks with a probabilistic approach. In: International Conference on Machine Learning, pp. 6727–6736. PMLR (2019)
204. Fazlyab, M., Morari, M., Pappas, G.J.: Probabilistic verification and reachability analysis of neural networks via semidefinite programming. In: 2019 IEEE 58th Conference on Decision and Control (CDC), pp. 2726–2731. IEEE (2019)
205. Wang, Y.S., Weng, T.W., Daniel, L.: Verification of neural network control policy under persistent adversarial perturbation (2019). arXiv preprint arXiv:1908.06353
206. Wang, B., Shi, Z., Osher, S.: Resnets ensemble via the feynman-kac formalism to improve natural and robust accuracies. Adv. Neural Inf. Process. Syst. **32** (2019)
207. Chiang, P.Y., Ni, R., Abdelkader, A., Zhu, C., Studor, C., Goldstein, T.: Certified defenses for adversarial patches. In: International Conference on Learning Representations (2020)
208. Latorre, F., Rolland. P., Cevher, V.: Lipschitz constant estimation of neural networks via sparse polynomial optimization. In: International Conference on Learning Representations (2020)
209. Jordan, M., Dimakis, A.G.: Exactly computing the local lipschitz constant of relu networks. Adv. Neural Inf. Process. Syst. **33**, 7344–7353 (2020)
210. Dvijotham, K.D., Hayes, J., Balle, B., Kolter, Z., Qin, C., Gyorgy, A., Xiao, K., Gowal, S., Kohli, P.: A framework for robustness certification of smoothed classifiers using f-divergences. In: International Conference on Learning Representations (2020)
211. Zhang, B., Cai, T., Lu, Z., He, D., Wang, L.: Towards certifying l-infinity robustness using neural networks with l-inf-dist neurons. In: International Conference on Machine Learning, pp. 12368–12379. PMLR (2021)
212. Zhang, B., Jiang, D., He, D., Wang, L.: Boosting the certified robustness of l-infinity distance nets. In: International Conference on Learning Representations (2022)
213. Liu, H., Jia, J., Gong, N.Z.: Pointguard: provably robust 3d point cloud classification. In: Proceedings of the IEEE/CVF conference on Computer Vision and Pattern Recognition, pp. 6186–6195 (2021)
214. Zhang, H., Wang, S., Xu, K., Li, L., Li, B., Jana, S., Hsieh, C.J., Kolter, J.Z.: General cutting planes for bound-propagation-based neural network verification. Adv. Neural Inf. Process. Syst. **35**, 1656–1670 (2022)
215. Zhang, B., Jiang, D., He, D., Wang, L.: Rethinking lipschitz neural networks and certified robustness: a boolean function perspective. Adv. Neural Inf. Process. Syst. **35**, 19398–19413 (2022)
216. Schuchardt, J., Günnemann, S.: Invariance-aware randomized smoothing certificates. Adv. Neural. Inf. Process. Syst. **35**, 34302–34320 (2022)
217. Li, L., Zhang, J., Xie, T., Li, B.: Double sampling randomized smoothing. In: International Conference on Machine Learning, pp. 13163–13208. PMLR (2022)
218. Qi, X., Wang, J., Chen, Y., Shi, Y., Zhang, L.: Lipsformer: introducing lipschitz continuity to vision transformers. In: The Eleventh International Conference on Learning Representations (2023)
219. Alfarra, M., Bibi, A., Torr, P.H.S., Ghanem, B.: Data dependent randomized smoothing. In: Uncertainty in Artificial Intelligence, pp. 64–74. PMLR (2022)
220. Anderson, B.G., Pfrommer, S., Sojoudi, S.: Towards optimal randomized smoothing: a semi-infinite linear programming approach. In: International Conference on Machine Learning Workshop on Formal Verification of Machine Learning (2022)
221. Anderson, B.G., Sojoudi, S.: Certified robustness via locally biased randomized smoothing. In: Learning for Dynamics and Control Conference, pp. 207–220. PMLR (2022)
222. Pfrommer, S., Anderson, B.G., Sojoudi, S.: Projected randomized smoothing for certified adversarial robustness (2023). arXiv preprint arXiv:2309.13794

223. Tramer, F., Carlini, N., Brendel, W., Madry, A.: On adaptive attacks to adversarial example defenses. Adv. Neural Inf. Process. Syst. **33**, 1633–1645 (2020)
224. The mnist database of handwritten digits. http://yannlecun.com/exdb/mnist/
225. Stallkamp, J., Schlipsing, M., Salmen, J., Igel, C.: The German traffic sign recognition benchmark: a multi-class classification competition. In: The 2011 International Joint Conference on Neural Networks, pp. 1453–1460. IEEE (2011)
226. Bommasani, R., Hudson, D.A., Adeli, E., Altman, R., Arora, S., von Arx, S., Bernstein, M.S., Bohg, J., Bosselut, A., Brunskill, E., et al.: On the opportunities and risks of foundation models (2021). arXiv preprint arXiv:2108.07258
227. Lomuscio, A., Maganti, L.: An approach to reachability analysis for feed-forward relu neural networks (2017). arXiv preprint arXiv:1706.07351
228. Scott, C., Nowak, R.: A Neyman-Pearson approach to statistical learning. IEEE Trans. Inf. Theory **51**(11), 3806–3819 (2005)
229. Mohapatra, J., Ko, C.Y., Weng, L., Chen, P.Y., Liu, S., Daniel, L.: Hidden cost of randomized smoothing. In: International Conference on Artificial Intelligence and Statistics, pp. 4033–4041. PMLR (2021)
230. Pérez, J.C., Alfarra, M., Giancola, S., Ghanem, B., et al.: 3deformrs: certifying spatial deformations on point clouds. In: Proceedings of the IEEE/CVF Conference on Computer Vision and Pattern Recognition, pp. 15169–15179 (2022)
231. Redmon, J., Divvala, S., Girshick, R., Farhadi, A.: You only look once: unified, real-time object detection. In: Proceedings of the IEEE conference on computer vision and pattern recognition, pp. 779–788 (2016)
232. Carion, N., Massa, F., Synnaeve, G., Usunier, N., Kirillov, A., Zagoruyko, S.: End-to-end object detection with transformers. In: European Conference on Computer Vision, pp. 213–229. Springer (2020)
233. Xie, C., Wang, J., Zhang, Z., Zhou, Y., Xie, L., Yuille, A.: Adversarial examples for semantic segmentation and object detection. In: Proceedings of the IEEE International Conference on Computer Vision, pp. 1369–1378 (2017)
234. Wei, X., Liang, S., Chen, N., Cao, X.: Transferable adversarial attacks for image and video object detection. In: Proceedings of the 28th International Joint Conference on Artificial Intelligence, pp. 954–960 (2019)
235. Wu, S., Dai, T., Xia, S.T.: Dpattack: diffused patch attacks against universal object detection (2020). arXiv preprint arXiv:2010.11679
236. Zhao, Y., Yan, H., Wei, X.: Object hider: adversarial patch attack against object detectors (2020). arXiv preprint arXiv:2010.14974
237. Huang, H., Wang, Y., Chen, Z., Tang, Z., Zhang, W., Ma, K.K.: Rpattack: refined patch attack on general object detectors. In: 2021 IEEE International Conference on Multimedia and Expo (ICME), pp. 1–6. IEEE (2021)
238. Huang, H., Chen, Z., Chen, H., Wang, Y., Zhang, K.: T-sea: transfer-based self-ensemble attack on object detection. In: Proceedings of the IEEE/CVF Conference on Computer Vision and Pattern Recognition, pp. 20514–20523 (2023)
239. Zhang, H., Wang, J.: Towards adversarially robust object detection. In: Proceedings of the IEEE/CVF International Conference on Computer Vision, pp. 421–430 (2019)
240. Chen, P.C., Kung, B.H., Chen, J.C.: Class-aware robust adversarial training for object detection. In: Proceedings of the IEEE/CVF Conference on Computer Vision and Pattern Recognition, pp. 10420–10429 (2021)
241. Dong, Z., Wei, P., Lin, L.: Adversarially-aware robust object detector. In: European Conference on Computer Vision, pp. 297–313. Springer (2022)

242. Chen, L.C., Zhu, Y., Papandreou, G., Schroff, F., Adam, H.: Encoder-decoder with atrous separable convolution for semantic image segmentation. In: Proceedings of the European Conference on Computer Vision (ECCV), pp. 801–818 (2018)
243. Xie, E., Wang, W., Yu, Z., Anandkumar, A., Alvarez, J.M., Luo, P.: Segformer: simple and efficient design for semantic segmentation with transformers. Adv. Neural Inf. Process. Syst. **34**, 12077–12090 (2021)
244. Jain, J., Li, J., Chiu, M.T., Hassani, A., Orlov, N., Shi, H.: Oneformer: one transformer to rule universal image segmentation. In: Proceedings of the IEEE/CVF Conference on Computer Vision and Pattern Recognition, pp. 2989–2998 (2023)
245. Gu, J., Zhao, H., Tresp, V., Torr, P.H.S.: Segpgd: an effective and efficient adversarial attack for evaluating and boosting segmentation robustness. In: European Conference on Computer Vision, pp. 308–325. Springer (2022)
246. Yin, H., Wang, R., Liu, B., Yan, J.: On adversarial robustness of semantic segmentation models for automated driving. In: 2022 IEEE Intelligent Vehicles Symposium (IV), pp. 867–873. IEEE (2022)
247. Li, J., Shao, W., Wang, H.: Key challenges and Chinese solutions for sotif in intelligent connected vehicles. Engineering (2023)
248. Shi, S., Guo, C., Jiang, L., Wang, Z., Shi, J., Wang, X., Li, H.: Pv-rcnn: Point-voxel feature set abstraction for 3d object detection. In: Proceedings of the IEEE/CVF Conference on Computer Vision and Pattern Recognition, pp. 10529–10538 (2020)
249. Shi, S., Jiang, L., Deng, J., Wang, Z., Guo, C., Shi, J., Wang, X., Li, H.: Pv-rcnn++: point-voxel feature set abstraction with local vector representation for 3d object detection. Int. J. Comput. Vision **131**(2), 531–551 (2023)
250. Li, Y., Yu, A.W., Meng, T., Caine, B., Ngiam, J., Peng, D., Shen, J., Lu, Y., Zhou, D., Le, Q.V., et al.: Deepfusion: lidar-camera deep fusion for multi-modal 3d object detection. In: Proceedings of the IEEE/CVF Conference on Computer Vision and Pattern Recognition, pp. 17182–17191 (2022)
251. Bai, X., Hu, Z., Zhu, X., Huang, Q., Chen, Y., Fu, H., Tai, C.L.: Transfusion: robust lidar-camera fusion for 3d object detection with transformers. In: Proceedings of the IEEE/CVF Conference on Computer Vision and Pattern Recognition, pp. 1090–1099 (2022)
252. Caesar, H., Bankiti, V., Lang, A.H., Vora, S., Liong, V.E., Xu, Q., Krishnan, A., Pan, Y., Baldan, G., Beijbom, O.: nuscenes: a multimodal dataset for autonomous driving. In: Proceedings of the IEEE/CVF Conference on Computer Vision and Pattern Recognition, pp. 11621–11631 (2020)
253. Yang, K., Tsai, T., Yu, H., Panoff, M., Ho, T.Y., Jin, Y.: Robust roadside physical adversarial attack against deep learning in lidar perception modules. In: Proceedings of the 2021 ACM Asia Conference on Computer and Communications Security, pp. 349–362 (2021)
254. Wang, S., Wu, T., Chakrabarti, A., Vorobeychik, Y.: Adversarial robustness of deep sensor fusion models. In: Proceedings of the IEEE/CVF Winter Conference on Applications of Computer Vision, pp. 2387–2396 (2022)
255. Kong, L., Liu, Y., Li, X., Chen, R., Zhang, W., Ren, J., Pan, L., Chen, K., Liu, Z.: Robo3d: towards robust and reliable 3d perception against corruptions. In: Proceedings of the IEEE/CVF International Conference on Computer Vision, pp. 19994–20006 (2023)
256. Zhang, Y., Hou, J., Yuan, Y.: A comprehensive study of the robustness for lidar-based 3d object detectors against adversarial attacks. Int. J. Comput. Vis., 1–33 (2023)
257. Liu, Y., Kong, L., Cen, J., Chen, R., Zhang, W., Pan, L., Chen, K., Liu, Z.: Segment any point cloud sequences by distilling vision foundation models. Adv. Neural Inf. Process. Syst. **36** (2023)
258. Hu, Y., Yang, J., Chen, L., Li, K., Sima, C., Zhu, X., Chai, S., Du, S., Lin, T., Wang, W., et al.: Planning-oriented autonomous driving. In: Proceedings of the IEEE/CVF Conference on Computer Vision and Pattern Recognition, pp. 17853–17862 (2023)

259. Salzmann, T., Ivanovic, B., Chakravarty, P., Pavone, M.: Trajectron++: dynamically-feasible trajectory forecasting with heterogeneous data. In: Computer Vision-ECCV 2020: 16th European Conference, Glasgow, UK, 23–28 Aug 2020, Proceedings, Part XVIII 16, pp. 683–700. Springer (2020)
260. Yuan, Y., Weng, X., Ou, Y., Kitani, K.M.: Agentformer: agent-aware transformers for socio-temporal multi-agent forecasting. In: Proceedings of the IEEE/CVF International Conference on Computer Vision, pp. 9813–9823 (2021)
261. Cui, H., Radosavljevic, V., Chou, F.C., Lin, T.H., Nguyen, T., Huang, T.K., Schneider, J., Djuric, N.: Multimodal trajectory predictions for autonomous driving using deep convolutional networks. In: 2019 International Conference on Robotics and Automation (ICRA), pp. 2090–2096. IEEE (2019)
262. Cao, Y., Xiao, C., Anandkumar, A., Xu, D., Pavone, M.: Advdo: realistic adversarial attacks for trajectory prediction. In: European Conference on Computer Vision, pp. 36–52. Springer (2022)
263. Tan, K., Wang, J., Kantaros, Y.: Targeted adversarial attacks against neural network trajectory predictors. In: Learning for Dynamics and Control Conference, pp. 431–444. PMLR (2023)
264. Cao, Y., Xu, D., Weng, X., Mao, Z., Anandkumar, A., Xiao, C., Pavone, M.: Robust trajectory prediction against adversarial attacks. In: Conference on Robot Learning, pp. 128–137. PMLR (2023)
265. Duan, A., Wang, R., Cui, Y., He, P., Chen, L.: Causal robust trajectory prediction against adversarial attacks for autonomous vehicles. IEEE Internet Things J. (2023)
266. Bubeck, S., Sellke, M.: A universal law of robustness via isoperimetry. Adv. Neural. Inf. Process. Syst. **34**, 28811–28822 (2021)
267. Ye, S., Xu, K., Liu, S., Cheng, H., Lambrechts, J.H., Zhang, H., Zhou, A., Ma, K., Wang, Y., Lin, X.: Adversarial robustness versus model compression, or both? In: Proceedings of the IEEE/CVF International Conference on Computer Vision, pp. 111–120 (2019)
268. Smith, L., Gal, Y.: Understanding measures of uncertainty for adversarial example detection (2018). arXiv preprint arXiv:1803.08533
269. Tuna, O.F., Catak, F.O., Eskil, M.T.: Closeness and uncertainty aware adversarial examples detection in adversarial machine learning. Comput. Electr. Eng. **101**, 107986 (2022)
270. Schweighofer, K., Aichberger, L., Ielanskyi, M., Klambauer, G., Hochreiter, S.: Quantification of uncertainty with adversarial models. Adv. Neural Inf. Process. Syst. 36 (2023)
271. Kirillov, A., Mintun, E., Ravi, N., Mao, H., Rolland, C., Gustafson, L., Xiao, T., Whitehead, S., Berg, A.C., Lo, W.Y., et al.: Segment anything (2023). arXiv preprint arXiv:2304.02643
272. Gu, A., Dao, T.: Mamba: linear-time sequence modeling with selective state spaces (2023). arXiv preprint arXiv:2312.00752
273. Zhu, L., Liao, B., Zhang, Q., Wang, X., Liu, W., Wang, X.: Vision mamba: efficient visual representation learning with bidirectional state space model (2024). arXiv preprint arXiv:2401.09417
274. Gui, S., Wang, H., Yang, H., Yu, C., Wang, Z., Liu, J.: Model compression with adversarial robustness: a unified optimization framework. Adv. Neural Inf. Process. Syst. 32 (2019)
275. Xiao, Y., Liu, A., Zhang, T., Qin, H., Guo, J., Liu, X.: Robustmq: benchmarking robustness of quantized models. Vis. Intell. **1**(1), 30 (2023)

Stable Resampling Strategies for Radar-Based Dynamic Occupancy Grids

Max Peter Ronecker, Michael Stolz and Daniel Watzenig

Abstract

Dynamic Occupancy Grid algorithms are an essential part of modern perception systems for autonomous vehicles, allowing for a combined representation of static and dynamic environments. However, these algorithms are known to be computationally expensive and difficult to optimize. This paper presents a detailed analysis of a radar-based DOGM algorithm, focusing on computation time and identifying key computational bottlenecks, particularly in the particle filter's resampling step. Two variations of the resampling process are proposed, effectively reducing the mean and variance of computation times without compromising performance. These modifications result in a 7% decrease in average computation time and approximately 50% less variance for higher particle counts.

Keywords

Dynamic occupancy grid mapping • Object detection • Computation time • Autonomous vehicles

M. P. Ronecker (✉)
SETLabs Research GmbH, Munich, Germany

M. Stolz · D. Watzenig
Virtual Vehicle Research GmbH, Munich, Germany
e-mail: michael.stolz@v2c2.at

D. Watzenig
e-mail: daniel.watzenig@v2c2.at

D. Watzenig (ed.), *Recent Advances in Autonomous Vehicle Technology–Perception and Path Planning*, Sustainable Mobility & Energy,
https://doi.org/10.1007/978-3-032-01001-8_3

1 Introduction

Autonomous technologies, from autonomous vehicles to driverless trains, are becoming foundational to the advancement of modern transport systems. These systems rely on their ability to continuously assess and respond to their dynamic surroundings. The effectiveness and safety of these autonomous technologies largely depend on how accurately and efficiently they can perceive environmental changes, which in turn influences their decision-making processes.

A key component in the framework of these systems is Dynamic Occupancy Grid Mapping (DOGM). This crucial perception technique not only detects static and dynamic objects, but also identifies areas of free space within the environment. It enhances the traditional occupancy grid approach [8] by integrating a particle filter, which improves its capabilities to track the dynamics of a scene [5, 6]. Additionally, occupancy is classified as static (non-moving) or dynamic (moving). In Fig. 1, a visualization of a DOGM environment structure is provided.

This framework has been progressively enhanced in subsequent research [11, 14]. Additionally, combining measurements from lidar and radar sensors has been shown to improve the performance of the filter [10]. A Dempster-Shafer framework for DOGM was also developed [7, 15, 16], introduced in various research efforts [10, 21], and subsequently enhanced [9]. This framework has been extended to include tasks related to object detection and shape estimation [17–19, 22].

To classify a cell as static or dynamic within the dynamic occupancy grid, a substantial number of particles in each occupied cell is essential. However, this requirement leads to a significant computational cost, complicating the implementation of the DOGM algorithm to function in real-time [9]. In this chapter, resampling is identified as a computational bottleneck in DOGM algorithms. From this analysis, several resampling strategies for the particle filter are introduced to enhance computational efficiency. To assess the performance of these sampling strategies, the object detection capabilities of the DOGM algorithm are examined using the NuScenes validation dataset [4].

Contributions

In this chapter, an extensive analysis of the radar-based Dynamic Occupancy Grid Mapping (DOGM) algorithm is conducted, focusing on computation time and identifying computational bottlenecks.

1. Computation times at each step of the DOGM algorithm are analyzed, particularly the effects of varying particle and cell counts, to pinpoint computational constraints.
2. Resampling is identified as a key computational bottleneck in DOGM algorithms, and several resampling strategies are proposed to improve computational efficiency in the particle filter.
3. Additional computational bottlenecks are explored, with potential solutions and research directions suggested for addressing them.

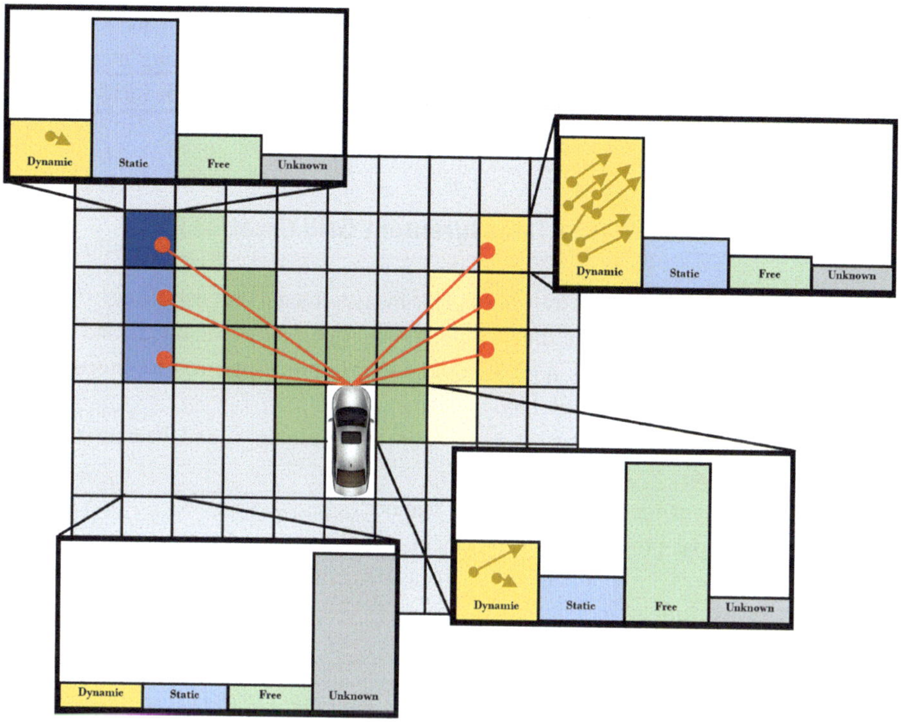

Fig. 1 Representation of the environment in a dynamic occupancy grid. Each grid cell is characterized by probabilities associated with four states: *unknown, free, static*, and *dynamic*. The dynamic properties within the grid are represented by a set of particles [13]

2 Dynamic Occupancy Grid Mapping

Subsequently, the algorithmic steps of the used radar-centric DOGM, based on [11–14], are introduced. These steps are performed repeatedly with each new measurement. An overview of the algorithm is provided in Fig. 2

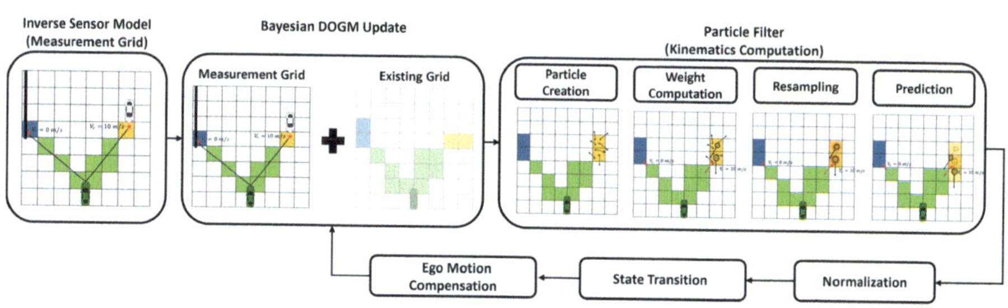

Fig. 2 An overview of the algorithmic steps of a radar-based DOGM algorithm [11–14]

The provided overview outlines a generalized structure although there are several variations. Some apply Dempster-Shafer theory or incorporate object tracking in a feedback loop [10, 17–19]. However, the core steps remain identical across different variants. Below, a brief description of each step is provided, as shown in Fig. 2.

2.1 Inverse Sensor Model (Measurement Grid Creation)

The Inverse Sensor Model (ISM) translates radar measurements into occupancy states. For a Dynamic Occupancy Grid, these states are unknown, free, static (occupied), and dynamic (occupied). As described in [12], a radar-centric approach is employed to directly derive these four states from the radial velocity.

The probability of state S at time $t+1$, given the measurement z_t and the map state m_{t-1} at time $t-1$, is defined as $P(S_{t+1}|z_t, m_{t-1})$.

In the case of a dynamic occupancy grid S can be $S_{unknown}$, S_{free}, S_{static} or $S_{dynamic}$.

Assuming the radar provides a measurement z_t indicating a detected object at a specific location together with a radial velocity measurements, the general form for updating according to Bayes' Rule is:

$$P(S_{t+1} = s|z_t, m_{t-1}) = \frac{P(z_t|S_{t+1} = s, m_{t-1})P(S_{t+1} = s|m_{t-1})}{P(z_t|m_{t-1})} \quad (1)$$

where:

$$\begin{aligned} P(z_t|S_{t+1} = s, m_{t-1}) &: \text{likelihood of observing } z_t \text{ given state } s, \\ P(S_{t+1} = s|m_{t-1}) &: \text{prior probability of state } s, \\ P(z_t|m_{t-1}) &: \text{normalization factor (evidence), computed as:} \end{aligned}$$

$$P(z_t|m_{t-1}) = \sum_{s\in\{unknown, free, static, dynamic\}} P(z_t|S_{t+1} = s, m_{t-1})P(S_{t+1} = s|m_{t-1}). \quad (2)$$

The Inverse Sensor Model (ISM) constructs a likelihood grid based on the latest sensor measurements. This grid is then used to update the prior occupancy grid. The environment is represented as a 2D Cartesian grid with equally sized square cells, where the ego vehicle is at the center.

For each state, the likelihood function $P(z_t|S_{t+1} = s, m_{t-1})$ needs to be defined based on the sensor characteristics and the environmental context. In general it can be described as:

- **Free Space (no object within the cell):**

$$P(z_t|S_{t+1} = free, m_{t-1}) = 1 - \text{occupancy probability}$$

- **Static Occupied (non-moving object within the cell):**

$$P(z_t|S_{t+1} = static, m_{t-1}) = \text{occupancy probability} \times \text{probability cell is not dynamic}$$

- **Dynamic Occupied (moving object within the cell):**

$$P(z_t|S_{t+1} = dynamic, m_{t-1}) = \text{occupancy probability} \times \text{probability cell is dynamic}$$

The formulation of probabilities depends on sensor accuracy, false-positive rate, and environmental factors. Static or dynamic classification is based on sensor data and deployment scenarios. An example of a radar-based dynamic occupancy grid using radial velocity is given in [12].

The "unknown" state is the default. It increases when a cell is not observed and decreases with new measurements. This follows a simpler model compared to the other states. A gradual increase in the unknown state reduces other states, slowly "forgetting" unobserved areas.

- **Unknown (cell not observed by measurements):**

$$P(S_{t+1} = unknown|z_t, m_{t-1}) = \text{A default value or a value decreasing with sensor information}$$

For this analysis, a simplified version of [12] is employed, using a Bresenham-like line algorithm [2] to determine the field of view while incorporating higher uncertainty in the angular direction.

2.2 Bayesian DOGM Update

This step updates the existing occupancy grid using data from the likelihood measurement grid. The measurement grid corresponds to the likelihood $P(z_t|S_{t+1} = s, m_{t-1})$, while the existing DOGM grid represents the prior $P(S_{t+1} = s|m_{t-1})$. The existing grid has already been predicted for the next time step by applying a motion model and state transition. The update can be performed using the method outlined in Eq. 1 or by employing a log-likelihood approach. Normalization, which involves dividing by the sum of the products, ensures that the probabilities remain within the range of 0 to 1. Below is an example for an update step of a single cell, with c being the cell index.

$$P(z_t|S_{t+1}=s, m_{t-1}) = \begin{bmatrix} P_{\text{meas},c}(unk.) \\ P_{\text{meas},c}(free) \\ P_{\text{meas},c}(stat.) \\ P_{\text{meas},c}(dyn.) \end{bmatrix} = \begin{bmatrix} 0.1 \\ 0.2 \\ 0.2 \\ 0.5 \end{bmatrix} \tag{3}$$

$$P(S_{t+1}=s|m_{t-1}) = \begin{bmatrix} P_{\text{dogm},c}(unk.) \\ P_{\text{dogm},c}(free) \\ P_{\text{dogm},c}(stat.) \\ P_{\text{dogm},c}(dyn.) \end{bmatrix} = \begin{bmatrix} 0.6 \\ 0.2 \\ 0.1 \\ 0.1 \end{bmatrix} \tag{4}$$

$$P(z_t|S_{t+1}=s, m_{t-1})P(S_{t+1}=s|m_{t-1}) = \begin{bmatrix} P_{\text{meas},c}(unk.)P_{\text{dogm},c}(unk) \\ P_{\text{meas},c}(free)P_{\text{dogm},c}(free) \\ P_{\text{meas},c}(stat.)P_{\text{dogm},c}(stat.) \\ P_{\text{meas},c}(dyn.)P_{\text{dogm},c}(dyn.) \end{bmatrix} = \begin{bmatrix} 0.06 \\ 0.04 \\ 0.02 \\ 0.05 \end{bmatrix} \tag{5}$$

$$P(z_t|m_{t-1}) = \sum_{s\in\{unk, free, stat., dyn.\}} P(z_t|S_{t+1}=s, m_{t-1})P(S_{t+1}=s|m_{t-1}) = 0.17 \tag{6}$$

$$P(S_{t+1}=s|z_t, m_{t-1}) = \begin{bmatrix} 0.06 \\ 0.04 \\ 0.02 \\ 0.05 \end{bmatrix} \frac{1}{0.17} = \begin{bmatrix} 0,35 \\ 0,24 \\ 0,12 \\ 0.29 \end{bmatrix} \tag{7}$$

As all values are stored in grid arrays this computation can be efficiently implemented using vectorized matrix operations.

2.3 Particle Filter (Kinematics Computation)

After determining the cell states, a particle filter (PF) is used to add precise velocity information. The four steps of the particle filter consist of *particle creation, weight computation, resampling* and *prediction.*

- *Particle Creation*: Particles are created only in cells where the probability of dynamic occupancy exceeds a specified threshold. Unlike Lidar-based approaches [9–11, 14, 17, 19, 21, 22], which generate particles in all occupied cells and classify them as dynamic or static later, the used method initiates particles solely in newly observed dynamic cells. These are integrated with existing particles. An example state vector for a particle is given by:

$$\mathbf{x}_p = \left[x_p, y_p, v_{p,x}, v_{p,y}, w_t\right]^T \tag{8}$$

Here, x_p and y_p are the coordinates within the grid, $v_{p,x}$ and $v_{p,y}$ are the velocity components, and w_t is the weight that quantifies the certainty in the particle's representation of the tracked object. The possible particle states can be constrained by considering surrounding measurements and applying simple heuristics [20].

- *Weight Computation*: After generating the particles, their weights are computed to reflect the likelihood that a particle's state matches the object being tracked, based on the latest measurements. Typically, this computation involves comparing the particle's velocity or position with the nearest measurements [10, 12, 13]. Particles that accurately represent the object's state receive higher weights. The dynamic state probability of each cell is distributed among the particles proportionally to their weights.
- *Resampling*: By sampling from the existing particles based on their weights and cell state probabilities, a new set of particles is created. This step filters out particles unlikely to represent the tracked objects, retaining those more likely to do so. Detailed information about the sampling procedure can be found in Sect. "Dynamic Occupancy Grid Mapping".
- *Prediction*: Finally, the particles are advanced to the next time step using a motion model; typically, a simple linear model is sufficient. During this step, the particles move into adjacent cells and transfer their respective dynamic probabilities, effectively predicting dynamic occupancy.

State Transition In the previous step, dynamic occupancy was predicted for the next time step. This prediction is now finalized by performing a state transition for all state probabilities. The state transition involves multiplying the cell state probability vector by a state transition matrix. A possible definition of the state transition matrix ST, based on [14], is as follows:

$$ST = \begin{pmatrix} 1 & 0.1 & 0.1 & 0.05 \\ 0 & 0.9 & 0 & 0 \\ 0 & 0 & 1 - P(v \neq 0) & P(v = 0) \\ 0 & 0 & P(v \neq 0) & 1 - P(v = 0) \end{pmatrix} \tag{9}$$

In this implementation, static and dynamic probabilities shift according to the cell's velocity. Probabilities are then normalized, accounting for the dynamic probabilities from the previous step, and particle weights are redistributed according to the new states.

2.4 Ego-Motion Compensation

Considering the ego-motion of the vehicle and its fixed central position as the reference point, the grid must be adjusted accordingly. This involves shifting the grid to align with the vehicle's movement, specifically by Δx and Δy in the x-y directions, while keeping the grid's orientation fixed. This allows the vehicle to rotate within the grid without the need to rotate the grid itself, ensuring improved spatial consistency between time steps [20].

3 Runtime Analysis

In this section, the runtime of the DOGM algorithm introduced in the previous chapter is thoroughly analyzed. The goal is to identify computational bottlenecks and understand the challenges and opportunities for optimization. The algorithm's complexity often necessitates sophisticated parallelization for real-time applications [9], requiring substantial engineering effort. Notably, many researchers in autonomous driving still employ CPUs for their non-machine learning algorithms. Consequently, the evaluations are performed on a CPU, without parallelization or GPU support. The implementation is in Python, focusing on efficiency and leveraging vectorization for optimal performance. Specifically, Python 3.10 and NumPy were used for most of the implementation.

The computational demand of the algorithm is primarily influenced by the number of cells and particles. The analysis, therefore, focuses on quantifying the impact of these parameters.

3.1 Quantifying the Impact of Particle Count

In the initial evaluation, a 100 m × 100 m occupancy grid with a fixed particle size was used. Three scenarios were tested with increasing numbers of particles: 62,500; 125,000; and 250,000. Unlike the original algorithm, where the number of particles varies with scene dynamics, the particle count was kept constant to better assess its impact on computation time. The results are shown in Fig. 3.

The data shows that the Particle Filter (PF) step requires the most significant amount of time and is heavily affected by an increasing number of particles. Within the PF steps, resampling (PF-Resampling) and secondly weight computation (PF-Weight) demand the longest computation time, a result that aligns with expectations given their direct involvement in modifying the particle set. Conversely, grid-related steps such as Initialization, Sensor Model (ISM), Update, and Ego-Motion Compensation (EMC) remain relatively constant in time, as anticipated.

In the second step, the behavior of dynamic particle allocation is examined in a more challenging setting with a 0.1 m resolution, leading to a higher average particle count. The results for the particle filter steps are displayed in Fig. 4.

It is evident that resampling dominates the computation time of the particle filter. Therefore, improving the resampling process could significantly reduce the overall computation time.

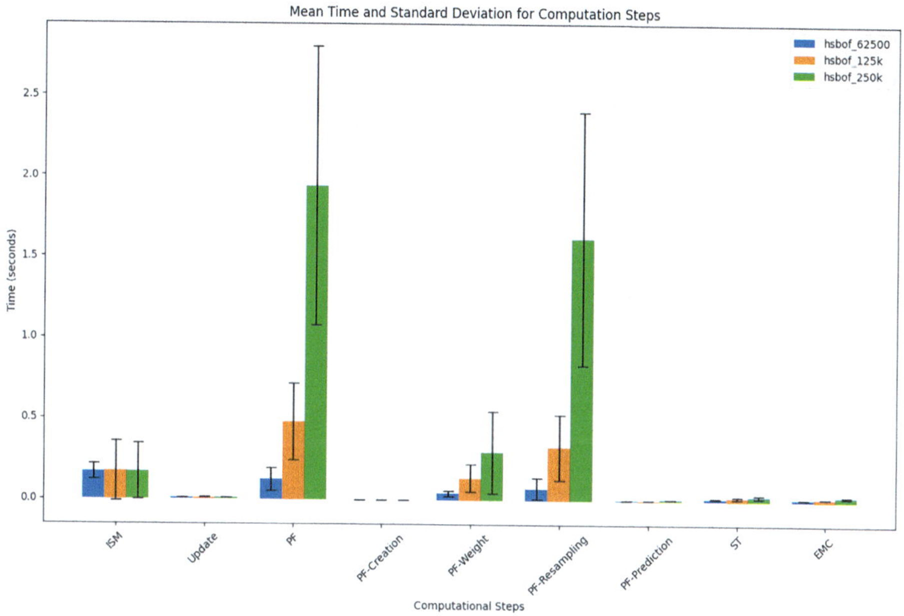

Fig. 3 Mean computation time and standard deviation across the various computation steps for different numbers of particles

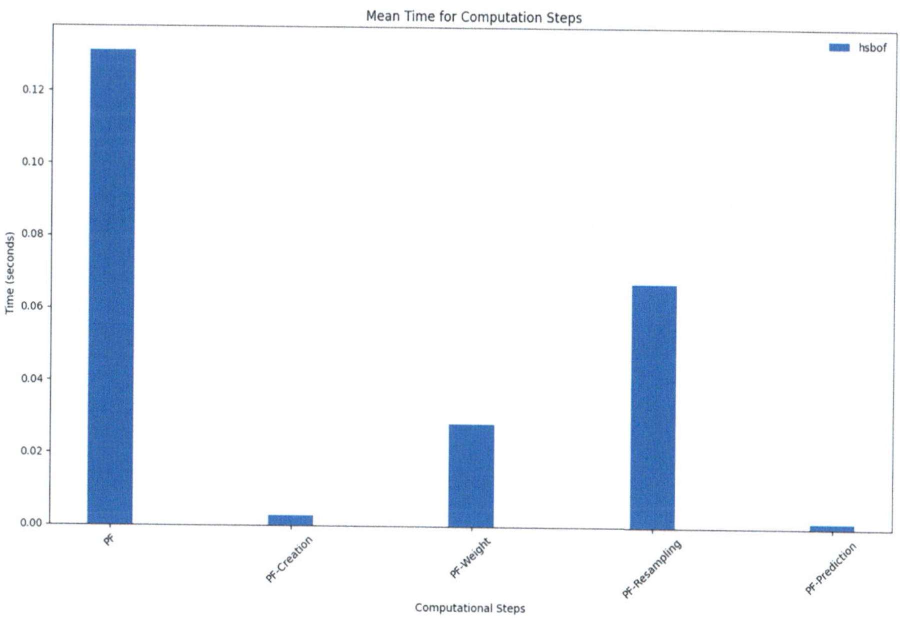

Fig. 4 Mean computation time and standard deviation across the various particle filter computation steps for a 0.1 grid and a dynamic particle allocation

3.2 Quantifying the Impact of Cell Count

Below, in Fig. 5, the computation times for various steps with an increasing number of cells are presented.

As expected, it is evident that the computation time for cell-related steps (such as ISM or state transition) increases with finer resolutions, resulting in a higher number of cells. Similarly, particle filter steps also require more computation time with an increased number of cells, as there are more cells to sample from. It can be seen that computation time and standard deviation for the Inverse Sensor Model more then doubles from 120 to 320 ms when going from a cell size of 0.5 m (40000 cells) to 0.1 m (1 million cells). For the state transition (ST) the exponential increase is visible as each cell has it's specific state transition. On the other side the particle filter related steps are largely independent of the cell count.

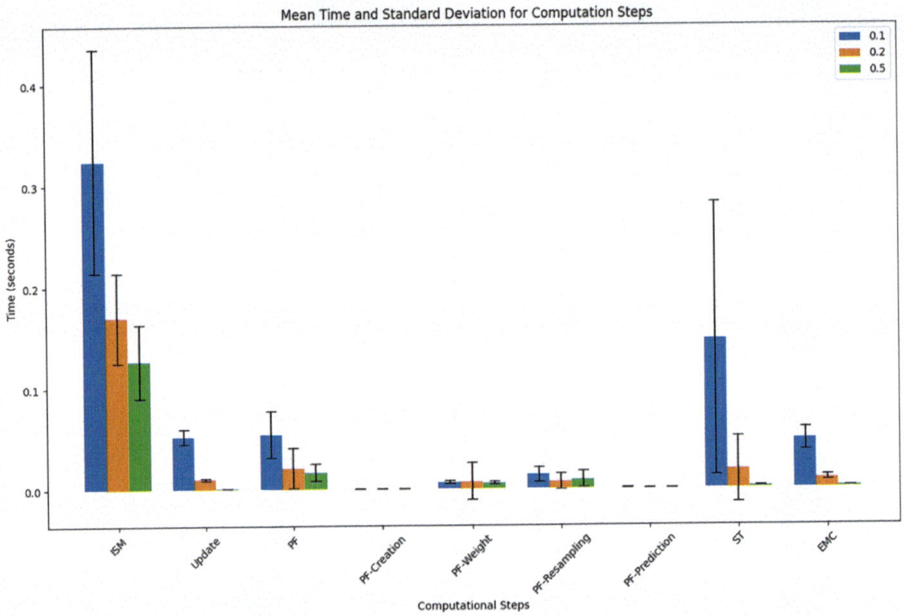

Fig. 5 Mean computation time and standard deviation across the various computation steps for different numbers of cells/cell sizes

3.3 Evaluation Results

The findings from the analysis of the DOGM algorithms computation time and challenges are summarized in the key points below:

- **Increased Particle Count Leads to Longer Computation Times**: Higher numbers of particles significantly extend the computation time for the Particle Filter, particularly during resampling steps. To address this, the number of necessary particles could be reduced as suggested in [12], or the efficiency of the Particle Filter, especially resampling, could be enhanced. Below, two modifications to the resampling step are proposed.
- **High Variability in Computation Time**: The computation time for all steps, especially those involving particles, shows high variance. This variability is attributed to the different compositions of the scene, which affect the number of dynamic cells and consequently impact resampling.
- **High Resolution Can Be Managed Efficiently**: Higher resolution does not affect computation time as significantly as particle count does. Computation time remains relatively consistent across different cell resolutions. This consistency is achieved by employing efficient array processing techniques such as vectorization. However, increased cell counts require more memory due to larger data structure sizes and typically result in more particles. To optimize this, selecting more efficient cell structures could be beneficial, as noted in [3, 23]. To enhance efficiency, choosing more effective cell structures could be beneficial, as highlighted in [3, 23]. Additionally, employing deep-learning-based ISMs [1, 24], which generate measurement grids directly from data, can also accelerate computation due to the integrated parallel processing and high efficiency of modern frameworks.

4 Efficient Sampling Strategies

In Sect. 3, resampling of the particles was identified as one of the major computational bottlenecks. This can be linked back to the applied resampling strategy of the "Hybrid-Sampling Bayesian Occupancy Filter" (HSBOF) [11], on which this approach is based. Based on this, two additional sampling strategies with potentially better computational properties are proposed. Below the implementation details are provided.

4.1 HSBOF Resampling Strategy

This resampling strategy, derived from the original HSBOF, employs a two-step process. First, cell indices are sampled according to their dynamic probabilities, with cells more likely to be dynamic sampled more frequently. Second, for each selected cell, a particle is chosen from those within the cell based on a sampling probability proportional to its weight. The steps of the HSBOF resampling procedure, based on [11], are shown in Algorithm 1.

Algorithm 1 HSBOF Resampling [11]

1: newParticles ← $\emptyset$
2: **for** $i = 1$ **to** numberParticles **do**
3: Draw cell index c by dynamic state probability $P(\text{dynamic} \mid c)$
4: Select particle p_i in cell c by weight distribution
5: Add p_i to newParticles
6: **end for**
7: Normalize weights of newParticles per cell

4.2 Vanilla Resampling

Under the term "Vanilla sampling," the standard resampling procedure of a particle filter is described. Here the resampling is done purely based on the weight of the particles independent of the cell probabilities. This approach eliminates the need for cell index sampling, significantly reducing computational complexity. In Algorithm 2 the steps are shown.

Algorithm 2 Vanilla Resampling

1: newParticles ← $\emptyset$
2: **for** $i = 1$ **to** numberParticles **do**
3: Select particle p_i based on weight distribution only
4: Add p_i to newParticles
5: **end for**
6: Normalize weights of newParticles per cell

Since cell probabilities are integrated into particle weights proportionally, separate cell index sampling may not be required. However, there is a concern that particles in cells with fewer particles might dominate due to receiving a larger portion of the cell's probability, unlike cells with many particles.

4.3 Cell-Based Sampling

As discussed in the previous chapter, Vanilla Resampling may allow single particles to dominate. To mitigate this, a modification is proposed by adding the overall cell probability to the weights of all particles within that cell. This adjustment creates an offset, ensuring that particles in cells with higher dynamic probabilities are sampled more frequently. The modified steps are shown in Algorithm 3.

Algorithm 3 Cell-based Resampling

newParticles $\leftarrow \emptyset$
for $i = 1$ **to** numberParticles **do**
  Calculate adjusted particle weight: $w_i^{'} = w_i + \text{cell\_probability}(p_i)$
end for
for $i = 1$ **to** numberParticles **do**
  Select particle p_i based on the adjusted weight $w_i^{'}$ distribution
  Add p_i to newParticles
end for
Normalize weights of newParticles per cell

Both the Vanilla and modified cell-based sampling methods avoid using for-loops to update and sample particles individually. Instead, they store cell locations, weights, and probabilities in array structures, allowing updates through matrix additions. Sampling is performed using a NumPy implementation.

5 Resampling Evaluation

The proposed resampling modification are compared with the original HSBOF-resampling by evaluating computation time and DOGM object detection performance. The analysis involved 150 diverse scenes from the nuscenes validation set, using a 100 m $\times$ 100 m grid at 0.1 m resolution, creating one million cells. Particle allocation varied dynamically with scene activity, and all resampling algorithms operated with identical particle creation parameters.

5.1 Computation Cost

First, the computation times of various resampling algorithms are examined. As demonstrated previously, performance strongly depends on the number of particles; therefore, all evaluations are based on particle count.

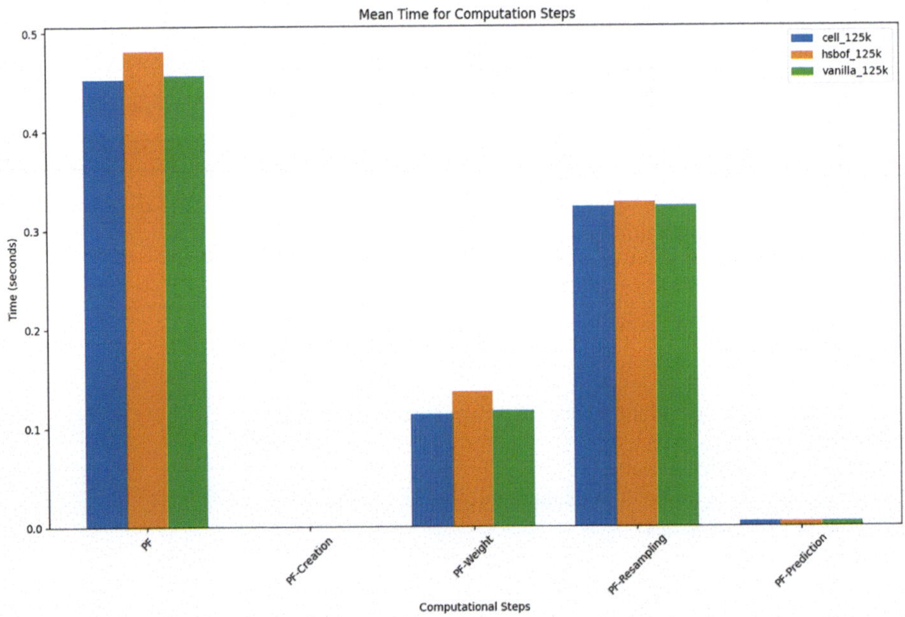

Fig. 6 Mean computation time and across the various particle filter computation steps for a grid with 0.2 m cell size and fixed 125k particles

Similar to the initial evaluation the new resampling algorithms are run on a grid with quadratic 0.2 m cells and a size of 100 m by 100 m, resulting in 250000 cells. 125000 is picked as fixed particle number (Fig. 6).

Overall, the computation time for the particle filter decreases from about 485 ms (hsbof) to about 450 ms for both vanilla and cell sampling, representing a 7% improvement. Interestingly, the resampling time remains quite constant, but the time required for weight computation decreases. This reduction can be attributed to the fact that fewer cells contain particles, thus fewer cells need to be accessed. However, performance remains consistent across all metrics, as shown in Sect. 5.2.

Additionally, the evaluation is run with dynamic particle allocation. Figure 7 displays the mean cycle times for three resampling algorithms—cell, hsbof, and vanilla—as particle counts increase for the dynamic allocation.

The vanilla and cell algorithms consistently exhibit the lowest resampling times, suggesting higher efficiency. Resampling time and variability increase with particle count across all algorithms, with both modifications outperforming HSBOF, particularly at higher particle counts. The difference between cell and vanilla resampling is minuscule. All resampling algorithms display relatively high standard deviations, indicating significant run-to-run variability, which underscores the impact of scene dynamics and computational factors on performance. However, both the cell-based and vanilla resampling methods consistently show

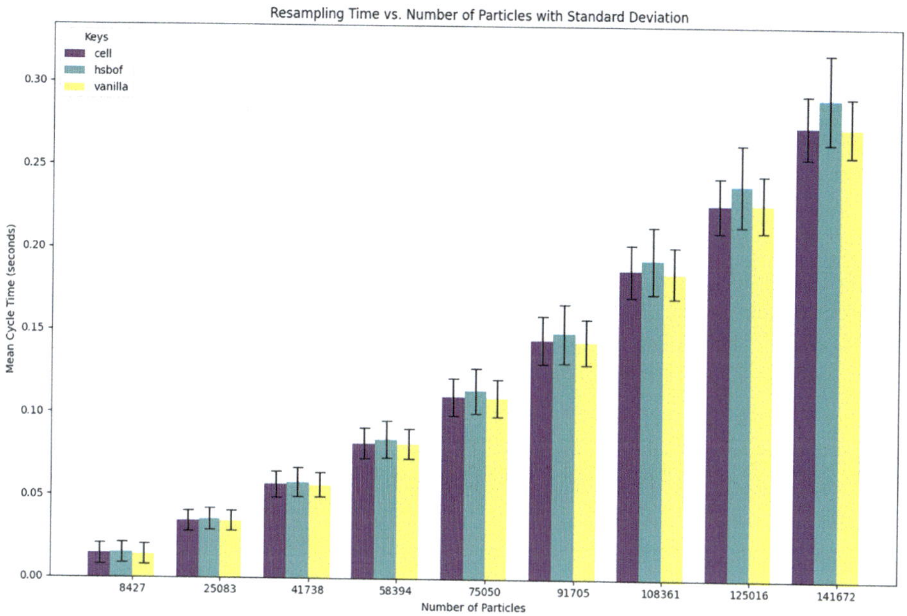

Fig. 7 Mean resampling time and standard deviation of the different algorithms with dynamic allocation depending on the particle count

lower deviations than the HSBOF algorithm. This discrepancy can be attributed to HSBOF's reliance on sampling from dynamic cells, where the number of cells varies depending on the scene. In scenarios with many dynamic patches, the pool of cells available for sampling increases.

Overall, the mean resampling time with the new algorithms remains relatively unchanged for smaller particle counts. For higher particle counts (>100k), the time can be reduced from 269 ms with HSBOF to 250 and 248 ms with cell and vanilla algorithms, marking an improvement of approximately 7%. Additionally, at these higher counts, the standard deviation improves significantly, dropping from 78 to 37/38 ms, which represents a more then 50% reduction. A lower standard deviation is beneficial as it ensures more consistent computation times. The improvement is consistent with the results for the static number of particles.

5.2 Object Detection Evaluation Results

In the previous section, it was demonstrated that the proposed modifications reduce resampling computation times. To ensure that these reductions do not compromise performance, object detection capabilities of the DOGM algorithm for the various resampling methods have been assessed. Specifically, focus is on tracking dynamic objects by evaluating the Average Precision for 4m (AP4) and velocity error Δv_{car} for all cars moving faster than 0.5 m/s in the NuScenes dataset. AP4 is one of the standard metrics for NuScenes, and 0.5 m/s is an empirically determined threshold for detecting dynamic objects in the current framework. Car is chosen as it's the most frequent class. In Table 1 the performance metrics for the static allocation are provided.

For this configuration, the overall performance is not good due to the simple ISM and suboptimal particle allocation. This setup primarily serves to validate computational improvements and ensure consistent object detection performance. The results for the dynamic allocation are shown in Table 2.

Object tracking performance is essentially identical across all resampling algorithms, with the AP value consistently around 38 and velocity error at approximately 1 m/s. This demonstrates that the same level of performance can be maintained with reduced computation time and variability. Notably, the lower standard deviation is advantageous in developing real-time systems, as it ensures more predictable computation times.

Table 1 Performance metrics for different resampling algorithms with static particle counts

Metric (dynamic)	HSBOF	Vanilla	Cell
$AP4_{car}$ (%)	17.5	18.0	18.9
Δv_{car} (m/s)	2.1	2.32	2.25
Particle filter time	485	449	454

Table 2 Performance metrics for different resampling algorithms with dynamic allocation

Metric (dynamic)	HSBOF	Vanilla	Cell
$AP4_{car}$ (%)	38.1	38.1	38.3
Δv_{car} (m/s)	1.01	1.03	1.01
Resampling time for >100k (ms)	269	248	250
Resampling standard deviation (ms)	78	37	38

6 Summary, Conclusion and Future Work

6.1 Summary

This paper examines the computational challenges in radar-based Dynamic Occupancy Grid Mapping (DOGM) algorithms, particularly the effect of particle and cell counts on efficiency. Resampling is identified as a major bottleneck, and two alternative methods are proposed: Vanilla Resampling and Cell-based Resampling. These methods reduce computation time by 7% and lower variance by 50% for higher particle counts while maintaining object detection performance. The findings help improve DOGM efficiency for real-time autonomous systems.

6.2 Conclusion

Resampling is a key factor in the computational cost of DOGM algorithms. Modifying the HSBOF resampling method results in lower computation time and more stable performance. These improvements are important for real-time applications where efficiency is critical. The results also show that high-resolution grids can be handled effectively with optimized data structures and vectorized computations. These modifications make radar-based DOGM more practical for autonomous systems by improving processing efficiency.

6.3 Future Work

Future research will focus on improving computational efficiency by integrating deep-learning-based inverse sensor models, which could replace traditional grid updates. A comparison between radar-based and multi-sensor DOGM approaches will also be conducted to evaluate the trade-offs between computation and performance. Additionally, further tests in more complex environments and larger datasets will help validate these resampling methods in real-world conditions.

Acknowledgements This work has received funding from the German Federal Ministry for Economic Affairs and Climate Action (BMWK) under grant agreement 19I21039A.

References

1. Bauer, D., Kuhnert, L., Eckstein, L.: Deep inverse sensor models as priors for evidential occupancy mapping. In: 2020 IEEE/RSJ International Conference on Intelligent Robots and Systems (IROS), pp. 6032–3067 (Oct. 2020). ISSN: 2153-0866

2. Bresenham, J.E.: Algorithm for computer control of a digital plotter. IBM Syst. J. **4**(1), 25–30 (1965)
3. Buerkle, C., Oboril, F., Jarquin, J., Scholl, K.-U.: Efficient dynamic occupancy grid mapping using non-uniform cell representation. In: 2020 IEEE Intelligent Vehicles Symposium (IV), pp. 1629–1634 (2020)
4. Caesar, H., Bankiti, V., Lang, A.H., Vora, S., Liong, V.E., Xu, Q., Krishnan, A., Pan, Y., Baldan, G., Beijbom, O.: nuscenes: a multimodal dataset for autonomous driving. In: CVPR (2020)
5. Coué, C., Pradalier, C., Laugier, C., Fraichard, T., Bessière, P.: Bayesian occupancy filtering for multitarget tracking: an automotive application. Int. J. Robot. Res. **25**(1), 19–30 (2006)
6. Danescu, R., Oniga, F., Nedevschi, S.: Modeling and tracking the driving environment with a particle-based occupancy grid. IEEE Trans. Intell. Transp. Syst. **12**(4), 1331–1342 (2011)
7. Dempster, A.P.: A Generalization of Bayesian Inference, pp. 73–104. Springer Berlin Heidelberg, Berlin, Heidelberg (2008)
8. Elfes, A.: Using occupancy grids for mobile robot perception and navigation. Computer **22**(6), 46–57 (1989)
9. Nuss, D., Reuter, S., Thom, M., Yuan, T., Krehl, G., Maile, M., Gern, A., Dietmayer, K.: A random finite set approach for dynamic occupancy grid maps with real-time application. Int. J. Robot. Res. **37**(8), 841–866 (2018)
10. Nuss, D., Yuan, T., Krehl, G., Stuebler, M., Reuter, S., Dietmayer, K.: Fusion of laser and radar sensor data with a sequential monte Carlo Bayesian occupancy filter. In: 2015 IEEE Intelligent Vehicles Symposium (IV), pp. 1074–1081 (2015)
11. Nègre, A., Rummelhard, L., Laugier, C.: Hybrid sampling Bayesian occupancy filter. In: 2014 IEEE Intelligent Vehicles Symposium Proceedings, pp. 1307–1312 (2014)
12. Ronecker, M.P., Diaz, X., Karner, M., Watzenig, D.: Dual-weight particle filter for radar-based dynamic Bayesian grid maps. In: 2024 IEEE International Conference on Robotics and Automation (ICRA) (2024)
13. Ronecker, M.P., Stolz, M., Watzenig, D.: Dual-weight particle filter for radar-based dynamic Bayesian grid maps. In: 2023 IEEE International Conference on Mobility, Operations, Services and Technologies (MOST), pp. 187–192 (2023)
14. Rummelhard, L., Nègre, A., Laugier, C.: Conditional monte Carlo dense occupancy tracker. In: 2015 IEEE 18th International Conference on Intelligent Transportation Systems, pp. 2485–2490 (2015)
15. Shafer, G.: A Mathematical Theory of Evidence. Princeton University Press (1976)
16. Smets, P.: The combination of evidence in the transferable belief model. IEEE Trans. Pattern Anal. Mach. Intell. **12**(5), 447–458 (1990)
17. Steyer, S., Lenk, C., Kellner, D., Tanzmeister, G., Wollherr, D.: Grid-based object tracking with nonlinear dynamic state and shape estimation. IEEE Trans. Intell. Transp. Syst. **21**(7), 2874–2893 (2020)
18. Steyer, S., Tanzmeister, G., Wollherr, D.: Object tracking based on evidential dynamic occupancy grids in urban environments. In: 2017 IEEE Intelligent Vehicles Symposium (IV), pp. 1064–1070 (2017)
19. Steyer, S., Tanzmeister, G., Wollherr, D.: Grid-based environment estimation using evidential mapping and particle tracking. IEEE Trans. Intell. Veh. **3**(3), 384–396 (2018)
20. Steyer, S.J.: Grid-Based Object Tracking. Ph.D. thesis, Technische Universität München, 2021
21. Tanzmeister, G., Thomas, J., Wollherr, D., Buss, M.: Grid-based mapping and tracking in dynamic environments using a uniform evidential environment representation. In: 2014 IEEE International Conference on Robotics and Automation (ICRA), pp. 6090–6095 (2014)
22. Tanzmeister, G., Wollherr, D.: Evidential grid-based tracking and mapping. IEEE Trans. Intell. Transp. Syst. **18**(6), 1454–1467 (2017)

23. Wellhausen, C., Clemens, J., Schill, K.: Efficient grid map data structures for autonomous driving in large-scale environments. In: 2021 IEEE International Intelligent Transportation Systems Conference (ITSC), pp. 2855–2862. IEEE Press (2021)
24. Weston, R., Cen, S., Newman, P., Posner, I.: Probably unknown: deep inverse sensor modelling radar. In: 2019 International Conference on Robotics and Automation (ICRA), pp. 5446–5452 (May 2019). ISSN: 2577-087X

Local Point Cloud Features for LiDAR Self-Supervised Representation Learning

Dušan Malić, Christian Fruhwirth-Reisinger and Horst Possegger

Abstract

Point clouds represent 3D spatial information using a set of discrete points relative to the sensor's origin. Such a representation captures global structure, but lacks local semantics, making it difficult to extract features rich in geometric detail. This aspect, however, remains largely overlooked in self-supervised pre-training tasks for LiDAR point clouds. To overcome this limitation, we propose a computationally efficient, yet highly effective approach: estimating local point cloud geometry from LiDAR data, specifically surface normals and curvature, and directly utilizing these features as network input. We demonstrate that incorporating our local geometric features improves the model's ability to learn a more discriminative latent representation through a self-supervised contrastive pre-training. Our evaluation demonstrates that a model pre-trained with our proposed local features outperforms a model pre-trained on standard global features for a LiDAR semantic segmentation task.

D. Malić (✉) · C. Fruhwirth-Reisinger · H. Possegger
Institute of Visual Computing, Graz University of Technology, Graz, Austria

Christian Doppler Laboratory for Embedded Machine Learning, Christian Doppler Laboratory, Vienna, Austria
e-mail: dusan.malic@tugraz.at

C. Fruhwirth-Reisinger
e-mail: reisinger@tugraz.at

H. Possegger
e-mail: possegger@tugraz.at

D. Watzenig (ed.), *Recent Advances in Autonomous Vehicle Technology–Perception and Path Planning*, Sustainable Mobility & Energy,
https://doi.org/10.1007/978-3-032-01001-8_4

Keywords

Domain generalisation • Surface normals • LiDAR point clouds • Local point features • Semantic segmentation

1 Introduction

Reliable perception models in autonomous vehicles are mostly based on LiDAR point clouds. A point cloud represents a discrete sampling of 3D space, characterized by a finite set of three-dimensional coordinates. Deep learning models directly process three-dimensional coordinates (i.e. global geometric features in the form of Cartesian coordinates) without preprocessing. Figure 1 illustrates that such common approach biases perception models towards positional information. Ideally, a model should consistently encode a specific point regardless of its position, since the inherent semantics of the point does not change based on its location. However, models trained on standard global features encode the same point differently if the entire point cloud undergoes a global transformation (e.g. translation, rotation, etc.).

Contrastive self-supervised pre-training approaches in the image domain [8, 18, 20] aim to learn spatially close embeddings for two augmented versions of the same image. They operate under the assumption that the underlying object features (shape, texture, etc.) are constant, regardless of the augmentation. Similarly, self-supervised pre-training methods are being applied to LiDAR data (e.g. [34, 53, 57, 60]). These methods aim to leverage the same contrastive self-supervised paradigm on point clouds. While standard global geometric input representations have been widely used, we argue it is sub-optimal for these tasks. Instead, we propose to convert global input representation into a local one.

Several approaches have been proposed to mitigate the dependency on global geometric information in 3D perception tasks. These methods can be broadly categorized into: methods utilizing local descriptors [38, 41, 43], decoupling global information with neural architectures [7, 27, 61], and enriching point semantics with image features [2, 6, 11]. However, the aforementioned approaches have been primarily tailored for carefully curated CAD and indoor RGB-D datasets, such as [10, 52, 56]. LiDAR perception tasks pose distinct challenges that set them apart from these datasets. These unique challenges include non-uniform point cloud density and generally lower point cloud density, the presence of noise due to variations in material properties and weather conditions, a substantially larger search space, and differences in object distributions. This substantial discrepancy in data hinders the direct adaptation of CAD/indoor methods to outdoor LiDAR data, and vice versa [47].

LiDAR-camera fusion, although explored in prior works [31, 32], comes with a computational overhead associated with processing additional image data alongside point clouds. Likewise, highly specialized architectures (e.g. [42]) often entail significant computational demands, posing a critical challenge in the development of autonomous driving systems.

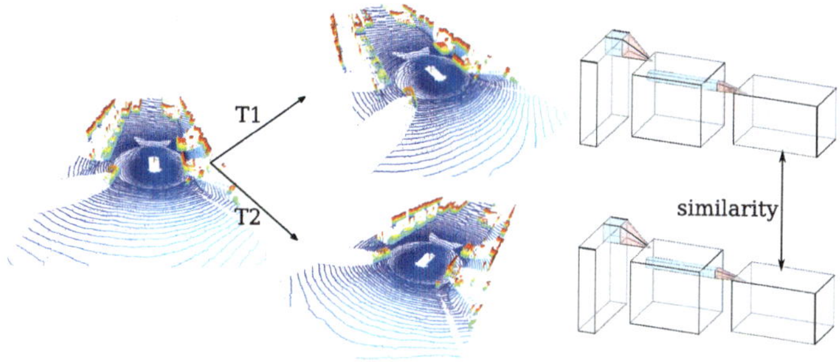

(a) Schematic representation of our experimental setup.

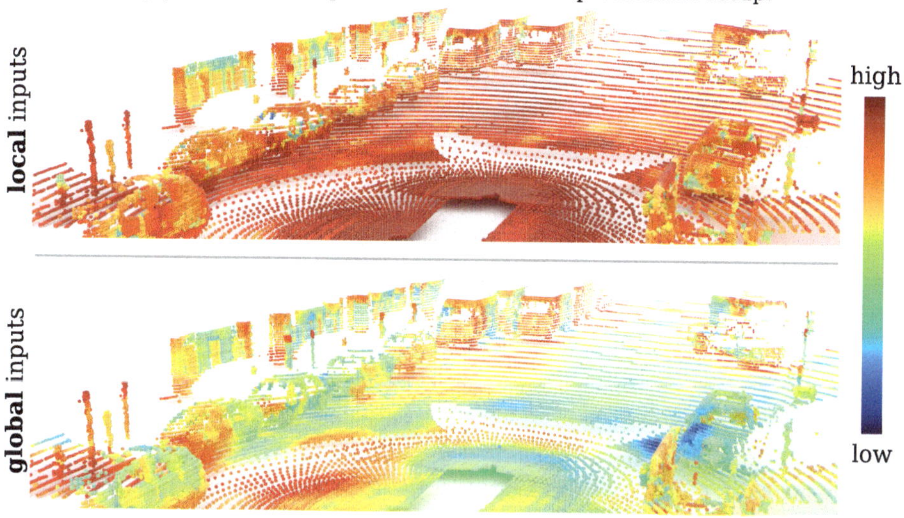

(b) Cosine similarity for our local and standard global input features.

Fig. 1 In this experiment, we augment a point cloud twice (T1 and T2) and feed it through an encoder in order to obtain per-point embeddings (depicted in **a**). We plot the cosine similarity between the embeddings of the two augmented versions of the same point cloud in **b** for our local (top) and standard global (bottom) inputs. Since the semantic information of each point does not change (only its location changes due to the augmentation), its embeddings should be highly similar, which we can achieve with our local input features (top, high per-point similarity). If the standard global input features are used (bottom), the same point is differently encoded due to the model's positional bias

To the best of our knowledge, there exists no line of work that explores local geometry for LiDAR perception tasks in autonomous driving.

In this work, we raise the following question: How can the inherent positional bias of LiDAR data be effectively mitigated for contrastive self-supervised pre-training tasks. To answer this, we investigate the utilization of different geometric features that leverage *local geometric information* within LiDAR point clouds. We propose converting the standard global input features into local geometric features, namely surface normal [23, 28, 33] and curvature [22, 45, 58], and utilizing them directly in both pre-training and downstream tasks. By decoupling global features from network inputs, we mitigate the positional bias and compel the model to learn a more semantically meaningful and discriminative feature representation. To empirically validate our findings, we evaluate the effectiveness of two recent contrastive self-supervised pre-training methods [34, 35] by pre-training a model on them and then fine-tuning it on a semantic segmentation task.

2 Related Work

SSRL for LiDAR point clouds Self-supervised Representation Learning (SSRL) aims to learn meaningful latent data representations in an unsupervised manner. Such representations can be effectively utilized to enhance the performance of diverse downstream tasks, such as object detection and semantic segmentation, or to reduce the labeling effort.

Masked autoencoders (MAEs) [19] utilize an encoder-decoder architecture during the pretraining phase to partially mask input point clouds and subsequently reconstruct the masked regions. The reconstruction is usually guided by occupancy prediction [21, 25, 46, 54]. Additionally, GeoMAE [46] highlights the effectiveness of incorporating an auxiliary task of surface normal and curvature reconstruction for achieving improved performance on downstream tasks.

Another line of work in SSRL adopts a contrastive learning framework [8] for representation learning. The network learns unsupervised object representations by enforcing latent space consistency of the same point cloud segment augmented twice [34, 35, 53]. SegContrast [34] partitions the entire point cloud into cohesive segments, subsequently employing these segments for positive and negative matching. TARL [35] utilizes a global reference frame to accumulate and segment sequential point clouds, enabling segment tracking over time. Positive pairs are established by using the same segment at different timestamps.

LiDAR semantic segmentation. A LiDAR semantic segmentation model aims to assign each LiDAR point to a predefined semantic label. In this context, we can roughly group existing methods based on the point cloud representation that they use.

Initial work in this domain utilized existing 2D Convolutional Neural Networks (CNNs) by applying them directly onto LiDAR range images [48, 49]. Building on this concept, PolarNet [59] utilizes 2D CNNs on a Bird's-Eye View (BEV) representation derived from a LiDAR point cloud. They demonstrate that a polar quantization scheme within the BEV representation outperforms a regular grid-based approach. However, information loss is an inevitable consequence of such approaches, due to the need for dimension compression.

Voxel-based methods, e.g. [9], address the limitations of 2D representations for point cloud data by directly processing the 3D data in voxel form. This approach, combined with efficient sparse 3D convolutions, significantly accelerates the model's inference time on 3D data. Reference [63] propose an improvement to standard cubic voxelisation with regular 3D convolutions by incorporating cylindrical voxelisation and asymmetrical 3D convolutions. This approach enhances the quantisation of sparse LiDAR point clouds with varying densities.

Depending on the voxel size, voxel-based approaches may produce overly smooth predictions for each individual point. Therefore, the PointNet [37] architecture operates directly on the unordered set of points. Similarly, a number of approaches directly apply transformer architectures to raw point clouds, utilizing varying receptive field mechanisms such as radial window [26], group vector [51] and point cloud serialization [50].

To achieve the best of both worlds, SPVConv [44] leverages both voxel-based and point-based architectures. Voxel-based approaches excel at capturing coarser features due to their larger receptive field, while point-based architectures encode fine-grained details.

Local point cloud features. The standard global representation is a convenient way to represent LiDAR point cloud data. However, even prior to the emergence of deep learning, there were compelling indications that local geometry, such as surface normal and curvature, could provide significant advantages [22, 62].

A surface normal describes a vector that is perpendicular to the surface at a given point. Traditional methods for surface normal estimation rely on Principal Component Analysis (PCA) [23], Voronoi diagrams [1] or the Hough transform [4]. These methods often necessitate meticulous parameter tuning and exhibit high sensitivity to noise and outliers. Therefore, they have often been superseded by learning-based methods, e.g. [28–30]. However, the drawback of such learning-based surface normal estimators is that they require ground truth data for supervision.

Surface curvature, characterized by the deviation of a surface from a flat plane, provides a quantitative measure of the local warping of a surface. Methods for estimating surface curvature include analyzing normal orientation changes [22], least squares fitting [58], and polyhedral approximation [45]. In addition, surface curvature can also be represented using a local umbrella structure [17].

Several approaches for CAD/indoor point clouds derive higher order geometric descriptors, using either surface normals or curvatures. Point Pair Features (PPF) [12–14] are anti-symmetric descriptors which encode relative distance and surface normals between two point pairs. Representative Surfaces (RepSurf) [38] encodes the local neighborhood as an umbrella-shaped surface, which is related to the umbrella curvature estimation [17]. While using higher-order descriptors can enhance local representation, we show that merely transforming global inputs into any local geometry leads to outstanding improvements with minimal pre-processing needed.

3 Local Point Cloud Features

LiDAR deep learning models represent input point clouds as a set of points denoted as $\mathcal{P} = \{\boldsymbol{p}_1, \ldots, \boldsymbol{p}_n\}$, where each point $\boldsymbol{p}_i = (x, y, z)$ denotes a global position in the three-dimensional space. This standard input representation, however, biases the detector towards learning global position cues, as shown in Fig. 1. To address these limitations, we propose a more suitable input representation for LiDAR point clouds that captures geometric context within the local neighborhood of the points: For each point in the point cloud $\mathcal{P}$, we estimate the surface normal $\mathbf{n}$ and curvature vector $\mathbf{c}$. A surface normal is a vector orthogonal to the tangent plane at a given point on a surface, while surface curvature quantifies the surface's deviation from a local reference plane at that point. In contrast to the standard global coordinate representation, using the surface normal $\mathbf{n}$ and curvature vector $\mathbf{c}$ as inputs facilitates the model in learning a discriminative latent representation. This leads to significantly improved semantic segmentation results, as our detailed evaluations demonstrate. In the following, we explain how to compute these suitable local features (Fig. 2).

3.1 Surface Normal Estimation

Surface normal estimation is a well-studied problem, with a wide variety of hand-crafted, e.g. [1, 4, 23], and learning-based methods, e.g. [28–30], being proposed over the years. Our findings suggest that downstream LiDAR models are relatively insensitive to the specific

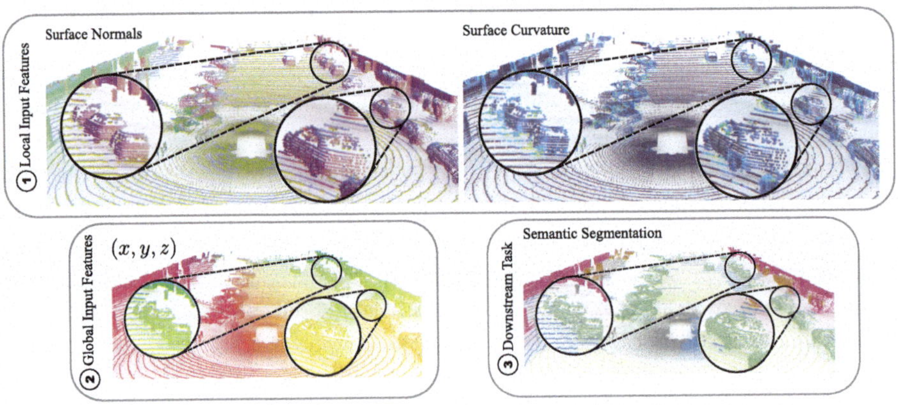

Fig. 2 Our local input features ① decouple the input representation from the global position within the frame. This property enables the network to effectively capture object characteristics independently of their spatial location, leading to more robust and generalizable representations compared to standard global input features ②. Our local input features are suitable for a variety of downstream tasks ③, which we demonstrate for semantic segmentation. To visualize these input features, we normalized their values and mapped them linearly to the RGB color space

surface normal estimation method. Noisy estimates can be effectively handled by the models through large-scale training, provided that the surface normal estimation method itself achieves a reasonable accuracy. Therefore, we adopt the classical, yet efficient Principal Component Analysis (PCA) of a local neighborhood, following [23].

We approximate the problem of estimating the normal to a point on a surface by estimating the normal of a plane $\boldsymbol{\Phi}$ tangent to this surface. This approximation yields a least squares plane fitting problem [40] for a small local neighborhood $\mathcal{V}(\boldsymbol{p}_q)$ of a query point $\boldsymbol{p}_q$. The plane's center $\boldsymbol{o}$ and surface normal $\boldsymbol{n}$ are the least squares estimates of $\left(\boldsymbol{p}_i - \boldsymbol{o}\right) \cdot \boldsymbol{n} = 0$. By specifying $\boldsymbol{o} = \overline{\boldsymbol{p}} = \frac{1}{k} \sum_{i=1}^{k} \boldsymbol{p}_i$, the surface normal $\boldsymbol{n}$ is given as the principal component of $\mathcal{V}(\boldsymbol{p}_q)$. The covariance matrix

$$\Sigma = \frac{1}{k} \sum_{i=1}^{k} \omega_i \left(\boldsymbol{p}_i - \overline{\boldsymbol{p}}\right) \left(\boldsymbol{p}_i - \overline{\boldsymbol{p}}\right)^{\top}, \quad \Sigma \in \mathbb{R}^{3\times 3} \tag{1}$$

is symmetric and positive semi-definite with real-valued eigenvalues λ_i and the corresponding eigenvectors $\boldsymbol{v}_i$. The eigenvector $\boldsymbol{v}_1$ corresponding to the smallest eigenvalue $\lambda_1 \leq \lambda_2 \leq \lambda_3$ is an approximation of the surface normal $+\boldsymbol{n}$ or $-\boldsymbol{n}$. Due to the inherent ambiguity of the PCA estimates, the surface normal orientation may not be consistent. By orienting surface normals to a consistent viewpoint, such that $\boldsymbol{n}\left(\mathrm{v}_p - \boldsymbol{p}_i\right) > 0$, we can resolve the ambiguous orientation. In particular, we use the sensor origin as the constant viewpoint v_p.

The uneven distribution of LiDAR data points within a frame, however, poses a significant challenge in defining a local neighborhood of a point for surface normal estimation. Close to the sensor, regions are sampled very densely, whereas the point density decreases with the region's distance to the sensor. Selecting either a too large neighborhood in dense regions, or a too small neighborhood in sparse regions leads to noisy surface normal estimates. To mitigate this, we rely on a point's k-nearest neighbors (k-NN), weighted by their distance, as this allows us to reduce the influence of outliers [39]. In particular, the weight ω_i in Eq. (1) is computed as

$$\omega_i = \begin{cases} \exp\left(-\frac{d_i^2}{\mu^2}\right) & d_i > d_{\max} \\ 1 & \text{otherwise,} \end{cases} \tag{2}$$

where d_i represents the distance of a query point to its closest neighbor, μ represents the mean distance of the query point to all its neighbors, and $d_{\max}$ defines the outlier threshold. This allows us to specify a relatively large $k = 32$, which enables good estimates in both dense and sparse regions of typical LiDAR point clouds.

3.2 Principal Curvature Estimation

Analyzing surface normal variations localized within a specific region provides valuable insights into the principal curvature characteristics of the surface. Given a query point $\boldsymbol{p}_q$ with its neighborhood $\mathcal{V}(\boldsymbol{p}_q)$, we project the estimated surface normals $\{\boldsymbol{n}_1, \ldots, \boldsymbol{n}_k\}$ onto the tangent plane $\boldsymbol{\Phi}_q$ by

$$\mathrm{proj}_{\boldsymbol{\Phi}_q}(\boldsymbol{n}_i) = \boldsymbol{n}_i - \frac{\boldsymbol{n}_i \cdot \boldsymbol{n}_q}{||\boldsymbol{n}_q||^2} \cdot \boldsymbol{n}_q, \quad \forall i \in [1, \ldots, k]. \tag{3}$$

Equivalent to the surface normal estimation, we perform PCA of normal projections to obtain the principal curvature estimates. The principal curvature direction $\boldsymbol{c}$ is the eigenvector corresponding to the largest eigenvalue of the covariance matrix. The maximum κ_1 and minimum κ_2 principal curvatures are the largest and the smallest eigenvalue, respectively.

In general, the curvature can be represented in a wide variety of ways, such as Gaussian $K = \kappa_1\kappa_2$ and mean $\overline{H} = \frac{\kappa_1+\kappa_2}{2}$ pair $\{K, \overline{H}\}$ [15], or via the shape index $S_I = \frac{1}{2} - \frac{1}{\pi}\arctan\left(\frac{\kappa_1+\kappa_2}{\kappa_1-\kappa_2}\right)$ [24], etc. We omit these estimates from our local representation, since they are very susceptible to both noise [22] and scale [36, 55]. Instead, we solely utilize the curvature direction $\boldsymbol{c}$.

4 Experiments

To evaluate the performance of our local geometric features, we conducted an empirical study comparing SegContrast [34] and TARL [35], two popular contrastive self-supervised pre-training methods. We investigated the impact of pre-training with default global vs. our proposed local features on a downstream semantic segmentation task.

Network architecture. SegContrast and TARL leverage the MinkUNet [9] architecture for pretraining and downstream fine-tuning. MinkUNet combines the popular U-Net structure, known for its ability to capture both high-level and detailed features, with the Minkowski Engine,[1] a framework that allows for efficient processing of sparse 3D data. This combination enables MinkUNet to effectively extract meaningful information from point clouds.

SegContrast. The prevalent contrastive representation learning strategy for semantic segmentation of LiDAR point clouds [53, 60] aims to increase the feature space similarity between two augmented representations of the same object, while simultaneously reducing the similarity to all other objects. SegContrast [34] leverages class-agnostic point segments to generate positive and negative samples. Enforcing high similarity of features from the same segment poses a significant optimization challenge when using the standard global

[1] https://github.com/NVIDIA/MinkowskiEngine.

Table 1 Semantic segmentation results (mIoU scores) on the SemanticKITTI [3] validation split. We follow TARL [35] and SegContrast [34], and pre-train the model on the SemanticKITTI train split. The columns show the evaluation results for different percentages of training data used for fine-tuning. The experiment configurations $xyzi$ and nci represent the default setup, including the intensity channel. We use models with random initialization (w/o pre-training) as our baseline

Method	0.1%	1%	10%	50%	100%
W/o pre-training$_{xyzi}$	**29.35**	42.77	53.96	**58.27**	59.03
W/o pre-training$_{nci}$	28.31	**43.04**	**54.46**	57.94	**59.83**
SegContrast$_{xyzi}$	32.75	44.83	56.31	60.45	61.02
SegContrast$_{nci}$	**35.64**	**49.39**	**58.21**	**60.99**	**61.59**
TARL$_{xyzi}$	38.59	51.42	60.34	61.42	61.47
TARL$_{nci}$	**38.88**	**53.44**	**60.71**	**62.00**	**62.99**

representation and the segments are spatially distant, as shown by the feature correlation in Fig. 1.

By incorporating our local input features, we alleviate the sensitivity of the contrastive objective to the positional bias. The SegContrast model trained with our local input features significantly enhances the performance of the default SegContrast model, resulting in a gain of almost 5 mIoU points on both SemanticKITTI [3] (Table 1) and nuScenes [5, 16] (Table 2) benchmarks. Pre-training with our local input features improves the model without pre-training by over 7 and 9 mIoU points on SemanticKITTI (Table 1) and nuScenes (Table 2), respectively. In contrast, the respective improvements of the default SegContrast model are only 3 and 4 mIoU points. Likewise, in Fig. 3, our model trained with local features achieves

Table 2 Semantic segmentation results (mIoU scores) on the nuScenes [5, 16] validation split. We use pretrained SegContrast [34] and TARL [35] checkpoints, fine-tuned on the nuScenes *Mini* and *Full* training splits. We follow TARL, and use SemanticKITTI [3] for pretraining. We use models with random initialization (w/o pre-training) as our baseline

Method	Mini	Full
w/o pre-training$_{xyzi}$	26.94	66.03
w/o pre-training$_{nci}$	**27.12**	**66.53**
SegContrast$_{xyzi}$	31.27	67.70
SegContrast$_{nci}$	**36.27**	**67.98**
TARL$_{xyzi}$	39.36	68.26
TARL$_{nci}$	**43.60**	**68.76**

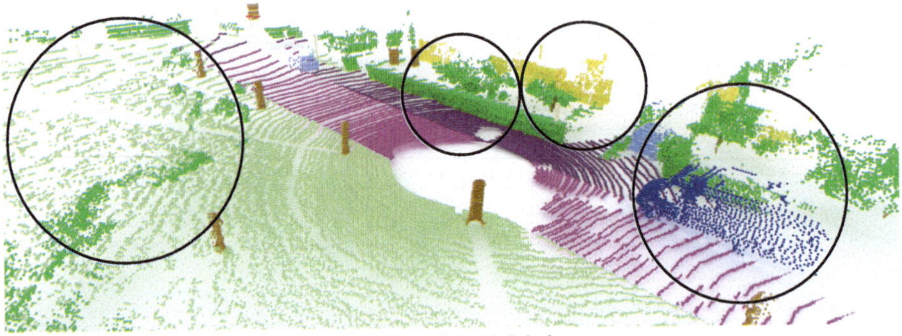

(a) Ground truth labels

(b) Results with SegContrast$_{xyzi}$

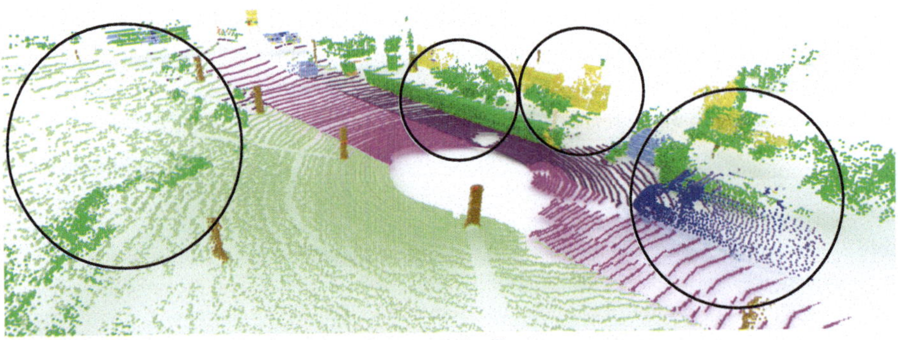

(c) Results with SegContrast$_{nci}$

Fig. 3 Qualitative comparison of ground truth labels and model predictions: **a** Ground truth label, **b** SegContrast$_{xyzi}$, and **c** SegContrast$_{nci}$ prediction on SemanticKITTI. Both models were pre-trained on the entire training split and subsequently fine-tuned on a subset of 1% of the data. We highlight the most noticeable distinctions observed between the two models' predictions

qualitatively better results than the baseline. This is particularly evident in regular planar regions, such as the van's side panels, bushes, and grass.

TARL. To corroborate our findings, we pre-train a TARL [35] model with our local input features on SemanticKITTI and fine-tune it on either SemanticKITTI or nuScenes. By incorporating our local input features, we are able to achieve substantial improvements in mIoU compared to TARL's default training approach, with gains of up to 2 (SemanticKITTI in Table 1) and 4 mIoU points (on nuScenes in Table 2). TARL's employment of an alternative positive matching strategy is likely responsible for its marginally lower relative improvement compared to SegContrast when trained with our local features. Instead of using the same segment augmented twice, TARL adopts a temporal aggregation approach that utilizes consecutive frames and considers the same segment from different time steps as a positive pair. Such an approach already partially mitigates the positional bias inherent to global features, due to the positive pairs being spatially close. However, our local input features facilitate the internal representation learning, resulting in further improvement gains.

4.1 Ablation Studies

The role of intensity. LiDAR intensity quantifies the strength of the reflected laser energy detected by a LiDAR sensor. This measure of the surface reflectivity is recorded as a scalar value per point, typically between 0 and 255, where higher values indicate stronger reflections. The strength of the reflected laser pulse is influenced by several factors, including the inherent reflectivity of the illuminated surface, the angle of incidence of the laser beam, the wavelength of the laser light, and the surface texture. For instance, highly reflective surfaces, such as metallic objects or water bodies, tend to generate stronger reflections, resulting in higher LiDAR intensity values. Conversely, surfaces with low reflectivity, such as vegetation or dark-colored materials, typically produce weaker reflections, yielding lower LiDAR intensity values.

In Fig. 4, we replicate our SSRL experiments, excluding the intensity channel from both the pre-training and fine-tuning phases. Relying solely on global input features for SegContrast pre-training consistently leads to subpar performance on the downstream task compared to the baseline model w/o pre-training. Conversely, pre-training and fine-tuning exclusively with our surface normal and curvature input features results in an enhancement of the baseline performance. The effectiveness of the local input features suggests that local information is essential for contrastive pre-training.

The default TARL architecture, when trained without the intensity channel, outperforms the baseline approach. While the specific component responsible for the TARL architecture's strong performance remains unclear (projection head, momentum update, or positive matching strategy), our findings consistently show improvements across all tested scenarios when using our local input features.

Extensive evaluation. Table 3 provides a comprehensive per-class evaluation on the nuScenes [5, 16] *Mini* training split. The TARL_{nci} model consistently demonstrates supe-

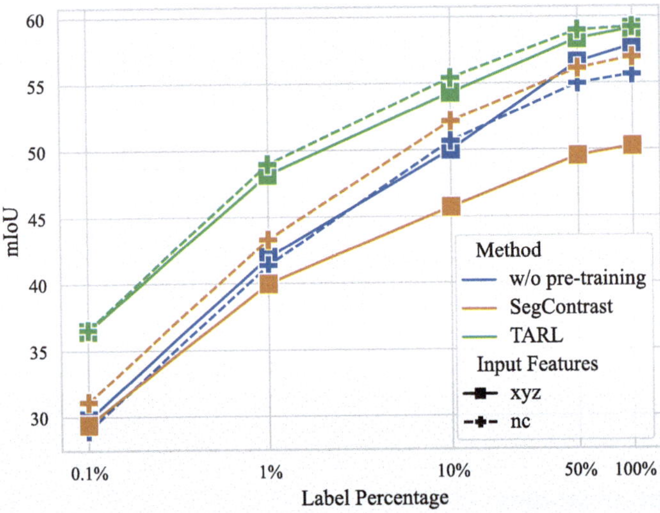

Fig. 4 mIoU scores for semantic segmentation on the SemanticKITTI [3] validation split. Here, we do not use the intensity channel for either pre-training or fine-tuning. We follow TARL [35] and SegContrast [34], and pre-train the model on the SemanticKITTI train split. We use models with random initialization (no pre-training) as our baseline. However, we do not use the intensity channel for either pre-training or fine-tuning. Evaluation results are shown for different percentages of training data used for fine-tuning. mIoU scores for semantic segmentation on the SemanticKITTI [3] validation split, excluding the intensity channel

rior performance compared to all other methods, achieving the highest mIoU scores and excelling in nearly all class-wise evaluations. Notably, our SegContrast$_{nci}$ model even outperforms TARL$_{xyzi}$ in approximately 25% of the evaluated object classes.

Runtime. To assess the scalability of our normal and curvature estimation, we perform ablation studies on three different datasets: Waymo, nuScenes, and KITTI (Field of View), with an average of 177K, 32K, and 20K points per frame, respectively. We perform our experiments locally on an Intel® Core™ i7-12700K CPU and an NVIDIA® RTX™ 4090 GPU.

Figure 5 shows the runtime for computing our local input features. The runtime scales linearly w.r.t the number of points and can be attributed to the combined complexity of identifying local neighborhoods in unstructured data and computing normals and curvatures from those neighborhoods. Leveraging the parallelization capabilities of GPUs, the GPU implementation achieves a performance improvement of nearly three orders of magnitude compared to the CPU counterpart. The runtime of the CPU implementation is also affected by the choice of k for the k-nearest neighbors (k-NN) algorithm. The computational complexity of estimating the local covariance matrix using Eq. (1) grows linearly with the value of k, introducing a significant bottleneck for CPU-based implementations. In contrast, GPUs

Table 3 Semantic segmentation results (mIoU scores) on the nuScenes [5, 16] validation split. We use pretrained SegContrast [34] and TARL [35] checkpoints, fine-tuned on the nuScenes *Mini* training split. We follow TARL, and use SemanticKITTI [3] for pretraining. We use models with random initialization (w/o pre-training) as our baseline

Method	mIoU	Barrier	Bicycle	Bus	Car	Construction Vehicle	Motorcycle	Pedestrian	Traffic Cone	Trailer	Truck	Driveable Surface	Other Flat	Sidewalk	Terrain	Manmade	Vegetation
W/o pre-training$_{xyzi}$	26.94	20.49	0.0	6.75	53.66	2.28	0.0	10.35	0.36	7.02	17.78	86.29	13.46	33.22	40.65	66.24	72.41
W/o pretraining$_{nci}$	27.12	15.77	00.	1.10	54.89	2.59	0.00	12.70	0.96	0.54	23.85	86.99	17.52	31.86	43.14	69.01	73.86
SegContrast$_{xyzi}$	31.27	23.50	0.0	10.73	71.70	5.10	0.22	28.80	8.47	8.16	18.72	84.39	16.55	33.27	45.10	69.89	75.66
SegContrast$_{nci}$	36.27	34.01	0	25.63	77.01	6.46	0.00	37.82	8.04	8.29	21.98	87.22	24.48	36.52	58.66	76.72	77.49
TARL$_{xyzi}$	39.36	24.59	0.0	21.95	74.27	6.97	**41.69**	42.70	11.18	8.18	37.50	88.54	19.39	**41.68**	55.64	76.36	**79.05**
TARL$_{nci}$	**43.60**	**34.04**	0.00	**36.25**	**78.02**	**14.06**	34.54	**47.17**	**16.75**	**8.36**	**51.38**	**88.56**	**31.34**	40.24	**60.26**	**77.88**	78.68

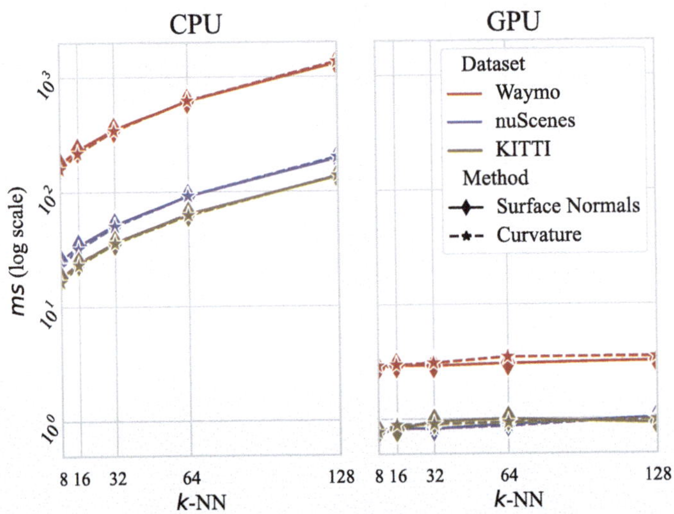

Fig. 5 Runtime for local feature computation on Waymo, nuScenes and KITTI (FOV) with an average of 177K, 32K and 20K points per frame, respectively

excel in handling such parallel computations, making them less susceptible to slowdown as k increases.

Surface normal and curvature estimation exhibits an average computational cost of 300 ms and 3 ms per Waymo point cloud on CPU and GPU, respectively. In practice, we employ the CPU implementation during the training phase of a model. This allows us to efficiently preprocess the input batches (on multiple CPU cores), since the bottleneck in training a model is the GPU. During inference, we utilize the GPU implementation to compute the local input features, exploiting the GPU's superior runtime performance.

5 Limitations and Conclusion

While our proposed local features introduce an additional pre-processing step, the associated runtime overhead is typically minimal. Even for very dense point clouds (e.g. Waymo), processing on the GPU typically takes only a few milliseconds, which is negligible in comparison to the semantic segmentation models. Despite its minimal cost, our evaluation highlights significant advantages of our local features such as reduced positional bias and superior self-supervised representation learning.

On the contrary, we demonstrated that the standard global input representation of LiDAR point clouds comes with limitations. Notably, positional bias can hinder the model's ability to learn internal representations in an unsupervised manner. Therefore, we introduced local geometric features, particularly surface normal and curvature, as a drop-in replacement for standard global features. Our proposed features are compatible with any network architecture and downstream task. Our work emphasizes that: (1) local geometric features have been vastly overlooked in LiDAR self-supervised representation learning tasks and (2) input data representation plays an equally crucial role as network architectures in shaping the effectiveness of deep learning-based models for LiDAR perception tasks. As demonstrated, our proposed local features can be easily integrated with any LiDAR perception model, leading to improved ability to learn effective internal representations in unsupervised settings.

Acknowledgements We gratefully acknowledge the financial support by the Austrian Federal Ministry for Digital and Economic Affairs, the National Foundation for Research, Technology and Development and the Christian Doppler Research Association. The presented experiments have been achieved using the Vienna Scientific Cluster. We would also like to thank the partners within the Christian Doppler Laboratory, especially AVL List GmbH, for the fruitful discussions and helpful practical insights.

References

1. Amenta, N., Bern, M.: Surface Reconstruction by Voronoi Filtering. In: Proc. SCG (1998)
2. Azuma, D., Miyanishi, T., Kurita, S., Kawanabe, M.: ScanQA: 3D Question Answering for Spatial Scene Understanding. In: Proc. CVPR (2022)
3. Behley, J., Garbade, M., Milioto, A., Quenzel, J., Behnke, S., Stachniss, C., Gall, J.: SemanticKITTI: A Dataset for Semantic Scene Understanding of LiDAR Sequences. In: Proc. ICCV (2019)
4. Boulch, A., Marlet, R.: Fast and robust normal estimation for point clouds with sharp features. In: CGFORUM. vol. 31, pp. 1765–1774 (2012)
5. Caesar, H., Bankiti, V., Lang, A.H., Vora, S., Liong, V.E., Xu, Q., Krishnan, A., Pan, Y., Baldan, G., Beijbom, O.: nuScenes: A multimodal dataset for autonomous driving. In: Proc. CVPR (2020)
6. Chen, D.Z., Chang, A.X., Nießner, M.: ScanRefer: 3D Object localization in RGB-D scans using natural language. In: Proc. ECCV (2020)
7. Chen, H., Liu, S., Chen, W., Li, H., Hill, R.: Equivariant point network for 3D point cloud analysis. In: Proc. CVPR (2021)
8. Chen, T., Kornblith, S., Norouzi, M., Hinton, G.: A simple framework for contrastive learning of visual representations. In: Proc. ICML (2020)
9. Choy, C., Gwak, J., Savarese, S.: 4D Spatio-Temporal ConvNets: Minkowski convolutional neural networks. In: Proc. CVPR (2019)
10. Dai, A., Chang, A.X., Savva, M., Halber, M., Funkhouser, T., Nießner, M.: ScanNet: Richly-annotated 3D Reconstructions of Indoor Scenes. In: Proc. CVPR (2017)
11. Dai, A., Nießner, M.: 3DMV: Joint 3D-Multi-View prediction for 3D semantic scene segmentation. In: Proc. ECCV (2018)
12. Deng, H., Birdal, T., Ilic, S.: PPF-FoldNet: Unsupervised learning of rotation invariant 3D local descriptors. In: Proc. ECCV (2018)
13. Deng, H., Birdal, T., Ilic, S.: PPFNet: Global Context Aware Local Features for Robust 3D Point Matching. In: Proc. CVPR (2018)
14. Drost, B., Ulrich, M., Navab, N., Ilic, S.: Model Globally, Match Locally: Efficient and Robust 3D Object Recognition. In: Proc. CVPR (2010)

15. Dyn, N., Hormann, K., Kim, S.J., Levin, D.: Optimizing 3D Triangulations Using Discrete Curvature Analysis. In: Proc. MMCS (2001)
16. Fong, W.K., Mohan, R., Hurtado, J.V., Zhou, L., Caesar, H., Beijbom, O., Valada, A.: Panoptic Nuscenes: A Large-Scale Benchmark for LiDAR Panoptic Segmentation and Tracking. RA-L **7**(2), 3795–3802 (2022)
17. Foorginejad, A., Khalili, K.: Umbrella curvature: A new curvature estimation method for point clouds. Procedia Technol. **12**, 347–352 (2014)
18. Grill, J.B., Strub, F., Altché, F., Tallec, C., Richemond, P., Buchatskaya, E., Doersch, C., Avila Pires, B., Guo, Z., Gheshlaghi Azar, M., et al.: Bootstrap your own latent a new approach to self-supervised learning. Proc, NeurIPS (2020)
19. He, K., Chen, X., Xie, S., Li, Y., Dollár, P., Girshick, R.: Masked autoencoders are scalable vision learners. In: Proc. CVPR (2022)
20. He, K., Fan, H., Wu, Y., Xie, S., Girshick, R.: Momentum contrast for unsupervised visual representation learning. In: Proc. CVPR (2020)
21. Hess, G., Jaxing, J., Svensson, E., Hagerman, D., Petersson, C., Svensson, L.: Masked autoencoder for self-supervised pre-training on Lidar point clouds. In: Proc. WACV Workshop (2023)
22. Hetzel, G., Leibe, B., Levi, P., Schiele, B.: 3D object recognition from range images using local feature histograms. In: Proc. CVPR (2001)
23. Hoppe, H., DeRose, T., Duchamp, T., McDonald, J., Stuetzle, W.: Surface reconstruction from unorganized points. In: SIGGRAPH (1992)
24. Koenderink, J.J., Van Doorn, A.J.: Surface shape and curvature scales. Image Vis. Comput. **10**(8), 557–564 (1992)
25. Krispel, G., Schinagl, D., Fruhwirth-Reisinger, C., Possegger, H., Bischof, H.: MAELi-Masked autoencoder for large-scale LiDAR point clouds. In: Proc. WACV (2024)
26. Lai, X., Chen, Y., Lu, F., Liu, J., Jia, J.: Spherical transformer for LiDAR-based 3D recognition. In: Proc. CVPR (2023)
27. Li, J., Chen, B.M., Lee, G.H.: SO-Net: Self-organizing network for point cloud analysis. In: Proc. CVPR (2018)
28. Li, Q., Feng, H., Shi, K., Gao, Y., Fang, Y., Liu, Y.S., Han, Z.: NeuralGF: Unsupervised point normal estimation by learning neural gradient function. In: Proc. NeurIPS (2023)
29. Li, Q., Feng, H., Shi, K., Gao, Y., Fang, Y., Liu, Y.S., Han, Z.: SHS-Net: Learning signed hyper surfaces for oriented normal estimation of point clouds. In: Proc. CVPR (2023)
30. Li, Q., Liu, Y.S., Cheng, J.S., Wang, C., Fang, Y., Han, Z.: HSurf-Net: Normal estimation for 3D point clouds by learning hyper surfaces. In: Proc. NeurIPS (2022)
31. Li, X., Ma, T., Hou, Y., Shi, B., Yang, Y., Liu, Y., Wu, X., Chen, Q., Li, Y., Qiao, Y., et al.: LoGoNet: Towards accurate 3D object detection with local-to-global cross-modal fusion. In: Proc. CVPR (2023)
32. Liu, Z., Tang, H., Amini, A., Yang, X., Mao, H., Rus, D.L., Han, S.: BEVFusion: Multi-Task multi-sensor fusion with unified bird's-eye view representation. In: Proc. ICRA (2023)
33. Mérigot, Q., Ovsjanikov, M., Guibas, L.J.: Voronoi-based curvature and feature estimation from point clouds. IEEE Trans. Vis. Comput. Graph. **17**(6), 743–756 (2010)
34. Nunes, L., Marcuzzi, R., Chen, X., Behley, J., Stachniss, C.: SegContrast: 3D point cloud feature representation learning through self-supervised segment discrimination. RA-L **7**(2), 2116–2123 (2022)
35. Nunes, L., Wiesmann, L., Marcuzzi, R., Chen, X., Behley, J., Stachniss, C.: Temporal consistent 3D LiDAR representation learning for semantic perception in autonomous driving. In: Proc. CVPR (2023)
36. Pauly, M., Keiser, R., Gross, M.: Multi-scale feature extraction on point-sampled surfaces. CGFORUM **22**(3), 281–289 (2003)

37. Qi, C.R., Su, H., Mo, K., Guibas, L.J.: PointNet: Deep learning on point sets for 3D classification and segmentation. In: Proc. CVPR (2017)
38. Ran, H., Liu, J., Wang, C.: Surface representation for point clouds. In: Proc. CVPR (2022)
39. Rusu, R.B., Marton, Z.C., Blodow, N., Dolha, M., Beetz, M.: Towards 3D point cloud based object maps for household environments. J. Robot. Auton. Syst. **56**(11), 927–941 (2008)
40. Shakarji, C.M.: Least-squares fitting algorithms of the NIST algorithm testing system. J. Res. Natl. Inst. Stand. Technol. **103**(6), 633 (1998)
41. Shen, Y., Feng, C., Yang, Y., Tian, D.: Mining point cloud local structures by kernel correlation and graph pooling. In: Proc. CVPR (2018)
42. Shi, W., Rajkumar, R.: Point-GNN: Graph neural network for 3D object detection in a point cloud. In: Proc. CVPR (2020)
43. Srivastava, S., Sharma, G.: Exploiting local geometry for feature and graph construction for better 3D point cloud processing with graph neural networks. In: Proc. ICRA (2021)
44. Tang, H., Liu, Z., Zhao, S., Lin, Y., Lin, J., Wang, H., Han, S.: Searching efficient 3D architectures with sparse point-voxel convolution. In: Proc. ECCV (2020)
45. Taubin, G.: Estimating the tensor of curvature of a surface from a polyhedral approximation. In: Proc. ICCV (1995)
46. Tian, X., Ran, H., Wang, Y., Zhao, H.: GeoMAE: Masked geometric target prediction for self-supervised point cloud pre-training. In: Proc. CVPR (2023)
47. Wang, Z., Li, Y.L., Chen, X., Zhao, H., Wang, S.: Uni3DETR: unified 3D detection transformer. In: Proc. NeurIPS (2023)
48. Wu, B., Wan, A., Yue, X., Keutzer, K.: SqueezeSeg: convolutional neural nets with recurrent CRF for real-time road-object segmentation from 3D LiDAR point cloud. In: Proc. ICRA (2018)
49. Wu, B., Zhou, X., Zhao, S., Yue, X., Keutzer, K.: SqueezeSegV2: Improved model structure and unsupervised domain adaptation for road-object segmentation from a LiDAR point cloud. In: Proc. ICRA (2019)
50. Wu, X., Jiang, L., Wang, P.S., Liu, Z., Liu, X., Qiao, Y., Ouyang, W., He, T., Zhao, H.: Point transformer V3: Simpler faster stronger. In: Proc. CVPR (2024)
51. Wu, X., Lao, Y., Jiang, L., Liu, X., Zhao, H.: Point Transformer V2: Grouped vector attention and partition-based pooling. In: Proc. NeurIPS (2022)
52. Wu, Z., Song, S., Khosla, A., Yu, F., Zhang, L., Tang, X., Xiao, J.: 3D ShapeNets: A deep representation for volumetric shapes. In: Proc. CVPR (2015)
53. Xie, S., Gu, J., Guo, D., Qi, C.R., Guibas, L., Litany, O.: PointContrast: Unsupervised pre-training for 3D point cloud understanding. In: Proc. ECCV (2020)
54. Yang, H., He, T., Liu, J., Chen, H., Wu, B., Lin, B., He, X., Ouyang, W.: GD-MAE: Generative decoder for MAE Pre-training on LiDAR point clouds. In: Proc. CVPR (2023)
55. Yang, Y.L., Lai, Y.K., Hu, S.M., Pottmann, H.: Robust principal curvatures on multiple scales. In: Proc. SGP (2006)
56. Yi, L., Kim, V.G., Ceylan, D., Shen, I.C., Yan, M., Su, H., Lu, C., Huang, Q., Sheffer, A., Guibas, L.: A scalable active framework for region annotation in 3D shape collections. ACM ToG **35**(6), 1–12 (2016)
57. Yin, J., Zhou, D., Zhang, L., Fang, J., Xu, C.Z., Shen, J., Wang, W.: ProposalContrast: Unsupervised pre-training for LiDAR-based 3D object detection. In: Proc. ECCV (2022)
58. Zhang, X., Li, H., Cheng, Z., et al.: Curvature estimation of 3D point cloud surfaces through the fitting of normal section curvatures. In: Proc. ASIAGRAPH (2008)
59. Zhang, Y., Zhou, Z., David, P., Yue, X., Xi, Z., Gong, B., Foroosh, H.: PolarNet: An improved grid representation for online LiDAR point clouds semantic segmentation. In: Proc. CVPR (2020)
60. Zhang, Z., Girdhar, R., Joulin, A., Misra, I.: Self-supervised pretraining of 3D features on any point-cloud. In: Proc. ICCV (2021)

61. Zhao, Y., Birdal, T., Lenssen, J.E., Menegatti, E., Guibas, L., Tombari, F.: Quaternion equivariant capsule networks for 3D point clouds. In: Proc. ECCV (2020)
62. Zhong, Y.: Intrinsic shape signatures: A shape descriptor for 3D object recognition. In: Proc. ICCV Workshop (2009)
63. Zhu, X., Zhou, H., Wang, T., Hong, F., Ma, Y., Li, W., Li, H., Lin, D.: Cylindrical and asymmetrical 3D convolution networks for LiDAR segmentation. In: Proc. CVPR (2021)

On Torque-Vectoring Control for the Obstacle Avoidance Scenario: A Reinforcement Learning Case Study

Carmine Caponio, Mario Mihalkov, Umberto Montanaro, Patrick Gruber, Aldo Sorniotti, and Zoltan Hankovszki

Abstract

Most of the research on vehicle stability control systems has dealt with feedback approaches, e.g., based on proportional-integral (PI), linear quadratic and model predictive control, which have been developed to generate direct yaw moments and longitudinal tire force distributions to compensate for large yaw rate errors and/or sideslip angle magnitudes, conditions in which the driver steering inputs are not able to significantly influence the cornering response. However, these strategies are still characterized by multiple challenges, such as the design complexity related to the necessity of considering tire nonlinearities and actuator-related dynamics and delays.

C. Caponio · M. Mihalkov · U. Montanaro · P. Gruber
School of Engineering, University of Surrey, Guildford, UK
e-mail: carmine.caponio@surrey.ac.uk

M. Mihalkov
e-mail: m.mihalkov@surrey.ac.uk

U. Montanaro
e-mail: u.montanaro@surrey.ac.uk

P. Gruber
e-mail: p.gruber@surrey.ac.uk

A. Sorniotti (✉)
Department of Mechanical and Aerospace Engineering, Politecnico di Torino, Torino, Italy
e-mail: aldo.sorniotti@polito.it

Z. Hankovszki
AVL List GmbH, Graz, Austria
e-mail: zoltan.hankovszki@avl.com

D. Watzenig (ed.), *Recent Advances in Autonomous Vehicle Technology–Perception and Path Planning*, Sustainable Mobility & Energy,
https://doi.org/10.1007/978-3-032-01001-8_5

To this purpose, in the last few years, artificial intelligence solutions have been proposed, among which reinforcement learning (RL) has shown promising results, thanks to its ability to solve the control task by learning through trial and error. In this field, only a few studies have evaluated RL applications for torque-vectoring control, with none of them considering the effect of the path tracking control layer in the context of automated vehicles. In this research, a real-time capable RL torque-vectoring controller for an automated vehicle is developed and assessed in obstacle avoidance maneuvers (according to the ISO 3888 standard), through an experimentally validated high-fidelity vehicle simulation model, including an automated driving control layer, as well as realistic tire behavior and actuation dynamics. In comparison with a rule-based PI algorithm, the results highlight the RL capability of flexibly tailoring the vehicle response depending on the selected reward function. The proposed RL agent's ability to learn control tasks through high-fidelity simulations could accelerate advanced control strategy design, and support benchmarking activities, by demonstrating optimal solutions for specific scenarios.

Keywords

Vehicle dynamics • Electric vehicles • Automated vehicles • Path tracking control • Torque-vectoring control • Machine learning • Reinforcement learning

1 Introduction

In the last 30 years, most of the academic and industrial research on vehicle stability control has been focused on deterministic approaches. Proportional-integral (PI) controllers, linear quadratic regulators (LQRs) and model predictive controllers (MPCs) have proven capable of improving the vehicle agility and stability characteristics through the application of direct yaw moment and braking actions, in the framework of yaw rate and/or sideslip angle control architectures [1–4]. In fact, in limit handling conditions, the steering inputs applied by the human or automated driver are not able to significantly influence the total yaw moment, so an external control action, e.g., in the form of a direct yaw moment, is required [1]. In conventional stability controllers, the direct yaw moment is applied only in critical conditions. However, modern electric vehicle (EV) architectures with multiple powertrains enable torque-vectoring (TV), i.e., the continuous and seamless actuation of direct yaw moment and total longitudinal tire force control [4].

As outlined in the review papers [5, 6], TV is a well-established technique that adjusts the longitudinal tire forces on each side of the vehicle to influence the cornering behavior, and enhance handling and stability also under extreme conditions. The rise of EVs has further emphasized the importance of TV, because of the presence of distributed powertrains, including multiple independently controlled motors. Additionally, electric machines can deliver the required torque more quickly and accurately than internal combustion engines, active differentials, or conventional hydraulic brakes.

Although the deterministic control methods have significantly enhanced the active safety performance, they are still characterized by several challenges [4, 7, 8], e.g.: (i) design complexity related to the necessity of embedding tire nonlinearities; (ii) difficulty in obtaining a suitable set of tuning parameters that ensure consistent performance across the broad range of potential scenarios; (iii) demanding state and parameter estimation requirements; and (iv) required robustness with respect to (w.r.t.) variations of vehicle and environmental parameters. Moreover, the real-time implementation of implicit MPC formulations is computationally demanding [7, 8].

To overcome (i)–(iv), artificial intelligence (AI) solutions have been recently proposed. In this context, reinforcement learning (RL) algorithms, which solve the control task by learning through trial and error [9–11], are especially promising. RL controllers (also known as *agents*) learn by collecting data (*states/observations*) from the *environment*, consisting of the nominal plant and its relevant disturbances. The targets imposed by the designer are reached by maximizing the so-called *reward function*, based on the numerical values coming from the environment, through multiple iterations, i.e., the *episodes*, see Fig. 1. The policy learned, i.e., the optimal behavior of the agent w.r.t. the specific input data, is stored in a look-up table or in a neural network (NN) [12, 13], which is the item deployed on the vehicle.

Among the variety of maneuvers proposed in the literature, the obstacle avoidance test at zero wheel torque demand, see the ISO 3888 standard, is frequently used to assess the

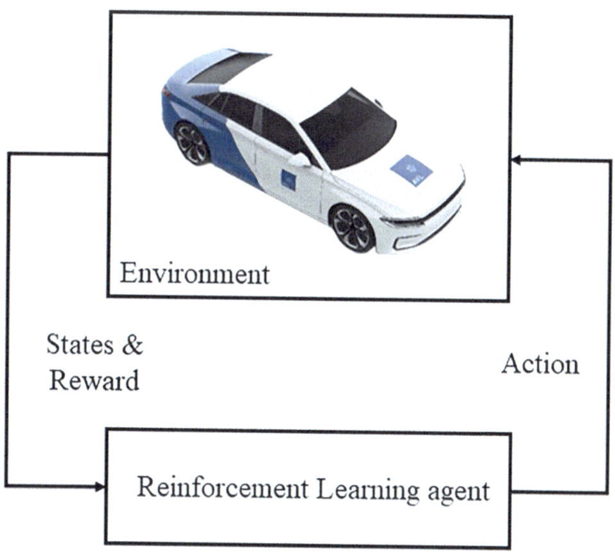

Fig. 1 Schematic of the agent-environment interaction during the training phase

vehicle dynamics performance. The objective is for the vehicle to complete the test from the highest possible entry speed, without hitting any of the cones indicating the boundaries of the course, while providing desirable response to the driver inputs. The test allows assessing the transient performance in terms of: (i) agility, i.e., the capability of performing swift changes of trajectory curvature to follow the desired path; and (ii) stability, i.e., the capability of operating in extreme transient conditions without generating significant sideslip angles at the rear axle [2]. In the literature, several controllers have been evaluated along this maneuver. For example, reference [1] presents an innovative MPC-based braking controller for connected vehicles with road preview. The algorithm slows down the vehicle only if its current speed is deemed safety critical w.r.t. the expected trajectory ahead, while allowing the cornering limit to be reached, and enabling the imposition of a soft constraint on the sideslip angle. In [2], LQR implementations generating reference steering and direct yaw moment actuations are applied to an automated vehicle with multiple electric powertrains, and are assessed through an experimentally validated model. The results emphasize the comparison between centralized and multi-layer control architectures.

In terms of RL implementations, in [7] an RL-based TV agent for combined vehicle stability and energy efficiency enhancement provides better yaw rate tracking as well as lower sideslip angle peaks and motor power losses, w.r.t. benchmarking MPC and LQR algorithms. However, the work in [7] only considers an open-loop test scenario, which does not appear to be very demanding from the vehicle dynamics viewpoint. A comparison between RL and MPC approaches to path tracking in automated vehicles, based on three levels of steering angle discretization (−5, 0, and 5°), is discussed in [14], which highlights that RL is more effective for a defined road scenario.

In summary, the following aspects remain underexplored in the literature: (i) the impact of incorporating the automated driving functionality in the simulation model to reflect the behavioral changes of the path tracking controller caused by the TV action; and (ii) a clear description and assessment of the reward function, which affects the agent training and the resulting policy learned by the controller. This research targets the gaps in the still limited RL vehicle dynamics control literature, through the following contributions:

- A real-time capable RL-based TV controller designed for the obstacle avoidance scenario according to the ISO standard.
- The assessment of the capability of the resulting agent to flexibly adapt its behavior according to the variation of the scaling factors of the reward function used in the training phase, for the same NN architecture and state space.
- A comparison with a benchmarking rule-based (RB) TV controller, across a comprehensive set of key performance indicators (KPIs), obtained through simulations with an experimentally validated high-fidelity simulation model, including realistic tire behavior and actuation dynamics.

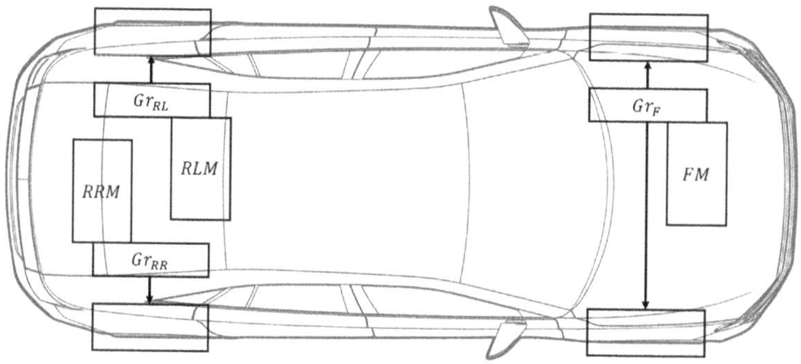

Fig. 2 Simplified schematic of the considered EV layout

The remainder is organized as follows: Sect. 2 presents the case study EV and its simulated control framework; Sect. 3 introduces the proposed control strategies; Sect. 4 discusses the case scenarios and simulation results; finally, Sect. 5 draws the main conclusions and outlines the future developments.

2 Case Study Vehicle and Simulation Framework

2.1 Case Study Electric Vehicle

The case study EV is a high-performance mass-production sedan. Figure 2 and Table 1 present the vehicle layout and its main parameters. The powertrain architecture consists of three on-board electric machines, controlled by the respective inverter, i.e., the front motor (FM in the figure), rear left motor (RLM), and rear right motor (RRM). These are coupled with single-speed transmissions connected to the wheels though half-shafts and constant velocity joints on the rear axle, while the front axle also includes an open differential. The three-motor architecture enables the generation of a corrective direct yaw moment through the variable distribution of the rear motor torque levels. A simulation model of the considered EV has been implemented in AVL VSM, a high-fidelity vehicle dynamics simulation package [15]. The model has been experimentally validated by the industrial participant involved in the activity. In the simulations of this study, the vehicle is controlled by an automated driving system, which defines the steering and torque demand inputs, where the latter can be corrected by the TV system.

2.2 Simulation and Control Framework

Figure 3 shows the implemented multi-layer control framework, including:

Table 1 Main vehicle parameters

Symbol	Description	Value
m	Vehicle mass	2476 kg
I_z	Yaw mass moment of inertia	3082 kg m^2
a_F	Front semi-wheelbase	1.485 m
a_R	Rear semi-wheelbase	1.485 m
h_{CG}	Center of gravity (CoG) height	0.458 m
t_F	Front track width	1.485 m
t_R	Rear track width	1.655 m
R_ω	Wheel radius	0.328 m
$T_{max,FM}$	Peak torque of front motor	425 N m
$T_{max,RjM}$	Peak torque of rear motor	160 N m
Gr_F	Gear ratio of the front drivetrain	8.6:1
Gr_{RjM}	Gear ratio of the rear drivetrain	16.7:1

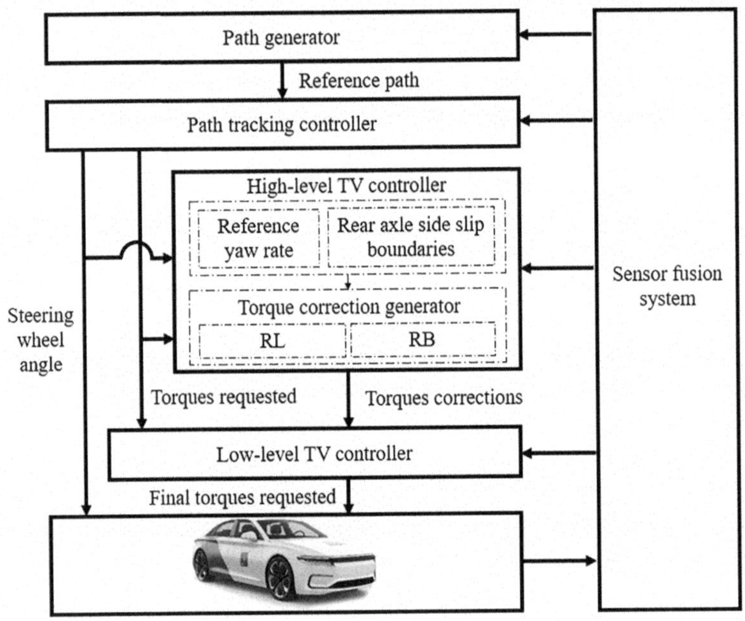

Fig. 3 Simplified schematic of the control architecture within the simulation framework

- The sensor fusion system, collecting the signals from the sensors located on the EV, and estimating/calculating the inputs required by the control blocks.
- The path generator, defining: (i) the reference longitudinal and lateral coordinates of the vehicle center of gravity (CoG), X_{ref} and Y_{ref}, in the inertial reference frame; ii) the reference curvature and heading angle, k_{ref} and ψ_{ref}; and (iii) the reference speed profile, V_{ref}, which is bypassed in the core part of the specific test of this study.
- The path tracking controller, which calculates the control actions for tracking the reference path. The block outputs the reference steering wheel angle, δ_{sw}, and the baseline values of the electric motor torque demands, $T_{ref,FM}$, $T_{ref,RLM}$, and $T_{ref,RRM}$, where the subscripts '*FM*', '*RLM*', and '*RRM*' indicate the relevant machine. In the implementation, the steering controller for trajectory tracking is the feedforward-feedback formulation from [2].
- The high-level TV controller, consisting of: (i) a generator of the reference yaw rate, $\dot{\psi}_{ref,TV}$. Similarly to conventional stability or TV control implementations, this is computed through a nonlinear map, as a function of steering wheel angle and vehicle speed. The look-up table of $\dot{\psi}_{ref,TV}$ was obtained through a set of ramp steer simulations in high friction conditions, each of them implying a slow steering input ramp at constant vehicle speed. 12 maneuvers were run in steps of 10 km/h, covering a speed range from 10 to 120 km/h [2, 3]. Appropriate first order dynamics are applied to the output of the static map. Only for the benchmarking RB controller, a correction is imposed when the sideslip angle exceeds a critical threshold of 5°, to obtain a soft constraint effect, according to the implementation in [3]. Based on the previous discussion, $\dot{\psi}_{ref,TV}$ is different from the time derivative of the reference heading angle ψ_{ref}; (ii) a function to compute the threshold value for the activation of the soft constraint on the rear axle slip angle, which is kept constant and equal to 5° in the analyses of this research; and (iii) the torque correction layer, including the RL agent and the benchmarking RB controller, which output the RLM and RRM torque corrections, $T_{corr,RLM}$ and $T_{corr,RRM}$.
- The low-level TV controller, which outputs the corrected electric motor torque demands, T_{FM}, T_{RLM}, and T_{RRM}, where $T_{FM} = T_{ref,FM}$, while the rear ones are given by:

$$T_{RjM} = T_{ref,RjM} - T_{corr,RjM}; \quad j \in [L, R] \tag{1}$$

3 Control Strategies

3.1 Reinforcement Learning Formulation

The formulations presented in this section and the remainder adopt the variables and sign conventions in Fig. 4. The state space s of the RL algorithm is the input vector of the resulting NN, i.e., the set of variables used by the agent to compute the control action. In this case, s is defined as:

$$s = \left[a_y, \delta_{sw}, V_x, V_y, \dot{\psi}, T_{app,RLM}, T_{app,RRM}, \beta, \int \beta dt, \Delta\dot{\psi}_{TV}, \int \Delta\dot{\psi}_{TV} dt\right] \tag{2}$$

where a_y is the lateral acceleration of the vehicle; V_x and V_y are the longitudinal and lateral components of the vehicle velocity; $\dot{\psi}$ is the yaw rate; β is the sideslip angle; t is time; $\Delta\dot{\psi}_{TV} = \dot{\psi} - \dot{\psi}_{ref,TV}$ is the yaw rate error; and $T_{app,RLM}$ and $T_{app,RRM}$ are the rear motor torque feedbacks, which consider the time delays and powertrain dynamics.

The action space a is defined as:

$$a = [T_{corr,RLM}, T_{corr,RRM}], \text{ with:} \\ -T_{max,RjM} \leq T_{corr,RjM} \leq T_{max,RjM} \tag{3}$$

The inequalities in (3) are hard constraints on the control inputs. They are meaningful only if $T_{ref,RLM} \approx T_{ref,RRM} \approx 0$, which is the case for the relevant part of the selected maneuver. To implement them, the output layer of the resulting NN is a tanh function, which is scaled by the peak value of the rear motor torque.

The final reward of the episode, R_{final}, is the sum of the reward function values, $R(i)$, for each time step of the episode:

$$R_{final} = \sum_{i=1}^{N_{episode}} R(i) = \sum_{i=1}^{N_{episode}} \left(R_{per}(i) + R_{cont}(i) + R_{cone}(i)\right) \tag{4}$$

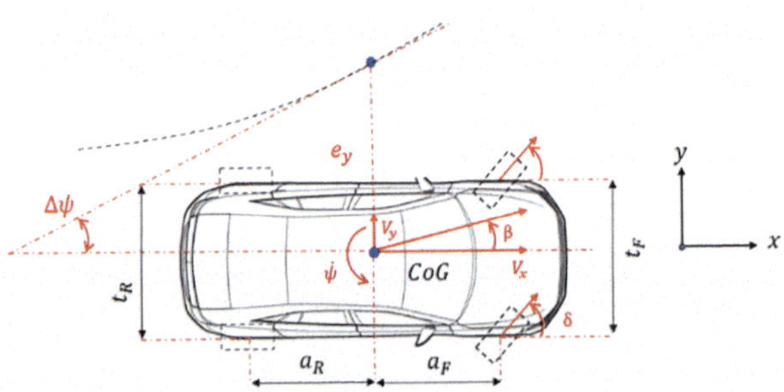

Symbol	Description
a_j	Semi-wheelbase
t_j	Track width
e_y	Lateral error
$\Delta\psi$	Heading error
δ	Wheel steering angle
V_x	Longitudinal vehicle speed
V_x	Lateral vehicle speed
β	Sideslip angle
$\dot{\psi}$	Yaw rate

Fig. 4 Simplified top view of the vehicle, showing the positive direction of the main relevant variables

where the integer values i and $N_{episode}$ indicate the current discretization step and the total number of steps within the episode; R_{per} is the sum of the rewards related to the performance targets the agent learns to achieve; R_{cont} is a term that depends on the control inputs, and limits the actuation effort; and R_{cone} penalizes the agent when the car hits a cone of the obstacle avoidance course. The R_{cone} contribution is realistic since the case study vehicle is characterized by driving automation, and in any case it is not part of s. Nevertheless, during the design phase, it was decided not to include the lateral and heading angle errors in the reward function, to keep the TV algorithm independent from the path tracking controller, consistently with the multi-layer and modular architecture recommended by the involved industrial partner.

The term R_{per} is defined as:

$$\begin{aligned}
R_{per} &= R_{\dot{V}_x} + k_{\alpha R} R_{\alpha R} + k_{\Delta\dot{\psi}} R_{\Delta\dot{\psi}}, \text{ with:} \\
R_{\dot{V}_x} &= -W_{\dot{V}_x}(|V_x(i) - V_x(i-1)|) \\
R_{\alpha R} &= \begin{cases} -W_{\alpha R}\log_{10}(\alpha_{R,th}), \; if\, |\alpha_R| < \alpha_{R,th} \\ -W_{\alpha R}\log_{10}(|\alpha_R|), \; if\, |\alpha_R| \geq \alpha_{R,th} \end{cases} \\
R_{\Delta\dot{\psi}} &= -W_{\Delta\dot{\psi}}\log_{10}\left(\left|\Delta\dot{\psi}_{TV}\right|\right)
\end{aligned} \tag{5}$$

where $R_{\dot{V}_x}$ is the reward contribution based on the variation of the longitudinal vehicle speed; $R_{\alpha R}$ is the reward associated to the control of the rear axle slip angle α_R; $R_{\Delta\dot{\psi}}$ is the yaw rate tracking reward; $W_{\dot{V}_x}$, $W_{\alpha R}$ and $W_{\Delta\dot{\psi}}$ are constant weights; and $k_{\alpha R}$ and $k_{\Delta\dot{\psi}}$ are tunable scaling factors to adjust priorities. The role of Eq. (5) can be summarized in the following points:

- $R_{\dot{V}_x}$ is designed to provide negative rewards to the agent. Since the vehicle always decelerates during the obstacle avoidance, this component guides the agent to avoid excessive vehicle deceleration and achieve maximum exit speed.
- $R_{\Delta\dot{\psi}}$ is described as a negative logarithmic function. The agent receives positive rewards for actions keeping the yaw rate error small, and thus the reduction of the yaw rate error is promoted.
- $R_{\alpha R}$ includes an 'if' condition to introduce a soft constraint on α_R. In fact, the agent learns to intervene only when α_R exceeds $\alpha_{R,th}$, set to 5° in the implementation, otherwise the reward remains constant. Limiting the reward for low values of α_R also reduces the training time as there is no convenience in exploring policies that shrink the sideslip angle.

R_{cont} is given by:

$$\begin{aligned}
R_{cont} &= R_{T_{corr}} + R_{\dot{T}_{corr}} + k_{\Delta T} R_{\Delta T}, \text{ with:} \\
R_{T_{corr}} &= -W_{T_{corr}}\left|T_{corr,RjM}\right|; j = L, R
\end{aligned}$$

$$\begin{aligned} R_{\dot{T}_{corr}} &= -W_{\dot{T}_{corr}} \left| \dot{T}_{corr,RjM} \right|; j = L, R \\ R_{\Delta T} &= -W_{\Delta T} |\Delta T| \end{aligned} \tag{6}$$

where $R_{T_{corr}}$, associated with the weight $W_{T_{corr}}$, penalizes the agent for each non-zero control input, to prevent unnecessary torque corrections; $R_{\dot{T}_{corr}}$, with its weight $W_{\dot{T}_{corr}}$, limits chattering on the control action; $R_{\Delta T}$, with its weight $W_{\Delta T}$, penalizes the agent when this does not achieve the automated driver torque demand, T_{ref}, i.e., it teaches the agent to generate a corrective direct yaw moment and follow T_{ref} instead of braking; $k_{\Delta T}$ is a tunable scaling factor; and the auxiliary variable ΔT is defined as:

$$\begin{aligned} \Delta T &= T_{ref} - T_{comm}, \text{ with:} \\ T_{ref} &= T_{ref,FM} + T_{ref,RLM} + T_{ref,RRM} \\ T_{comm} &= T_{FM} + T_{RLM} + T_{RRM} \end{aligned} \tag{7}$$

where T_{comm} is the total torque demand for the electric motors after the TV torque correction.

R_{cone} penalizes the agent when the vehicle crosses the specified lane boundaries:

$$\begin{aligned} R_{cone} &= -W_{cone}, \quad \text{if} F_{coll} = 1 \\ R_{cone} &= 0, \quad \text{if} F_{coll} = 0 \end{aligned} \tag{8}$$

where W_{cone} is a weight; and F_{coll} is a flag that becomes non-zero when the vehicle hits a cone, i.e., when $F_{coll} = 1$, the agent is heavily penalized, and the episode is stopped to reduce the training time.

The orders of magnitude of the weights of the reward terms have been chosen to: (i) allow the agent firstly to complete the maneuver by keeping the vehicle inside the boundaries defined by the cones. To this purpose, among the reward terms, R_{cone} is the one with the highest penalization; and (ii) avoid the agent slowing down and stopping the car, rather than stabilizing the vehicle and completing the maneuver.

Table 2 reports the values of the time step and weights of the considered RL agent. Differently from the weights, which are the same across the simulations of this study, the scaling factors were adjusted to implement different versions of the RL-based TV algorithm:

- Baseline, with the scaling factor settings defined by $k_{baseline} = \left[k_{\alpha_R,bs}, k_{\Delta\dot{\psi},bs}, k_{\Delta T,bs}\right]$.
- Performance, with reduced weights on R_{per} w.r.t. the baseline version, i.e., $k_{performance} = \left[0.1k_{\alpha_R,bs}, 0.1k_{\Delta\dot{\psi},bs}, k_{\Delta T,bs}\right]$. The target is to achieve the obstacle avoidance completion with higher exit speed.
- Safety, with increased scaling factors on R_{per} to prioritize α_R and $\Delta\dot{\psi}$ control, i.e., $k_{safety} = \left[2k_{\alpha_R,bs}, 2k_{\Delta\dot{\psi},bs}, 0.7k_{\Delta T,bs}\right]$.

Table 2 Main RL configuration parameters

Symbol	Description	Value
$T_{s,agent}$	Agent time step	10 ms
$W_{\dot{V}_x}$	Acceleration weight	1e6
$W_{\alpha R}$	Rear axle slip angle weight	1e4
$W_{\Delta\dot{\psi}}$	Reference yaw rate tracking weight	1e4
$W_{T_{corr}}$	Control action weight	0.3125
$W_{\dot{T}_{cor}}$	Control action derivative weight	0.3125
$W_{\Delta T}$	Total torque demand weight	1e4
W_{cone}	Weight penalizing cone collision	1e9

3.2 Agent Implementation

The RL algorithms were implemented with the reinforcement learning toolbox available in MATLAB 2023a [16]. The deep deterministic policy gradient (DDPG) actor-critic algorithm was used for the training phase, because of its good balance between sample efficiency and computational effort during training [12].

The actor network, i.e., the network that computes the torque corrections from the state space in (2), has two ReLu hidden layers with 100 neurons each, and one tanh output layer with two neurons. The critic network, i.e., the one that evaluates the performance of the actor during training, has two ReLu hidden layers with 100 neurons each for the state path, and two ReLu hidden layers with 30 and 100 neurons each for the action path. The training was considered to be completed after 5000 episodes. The controller sampling time is set to $T_{s,agent} = 10$ ms, which is a typical value for direct yaw moment control implementations within the automotive industry.

The real-time capability of the resulting trained agent was verified on the dSPACE MicroAutoBox II system in Fig. 5, with a 900 MHz processor and 16 Mb memory, which provided a maximum turnaround time of 0.065 ms.

3.3 Rule-Based Formulation

An RB strategy according to [3] was implemented as benchmarking controller, using a PI control action to generate the torque correction, $T_{corr,RB}$, which is then applied as follows:

$$\begin{aligned} T_{corr,RLM} &= T_{corr,RB} \\ T_{corr,RRM} &= -T_{corr,RB} \end{aligned} \tag{9}$$

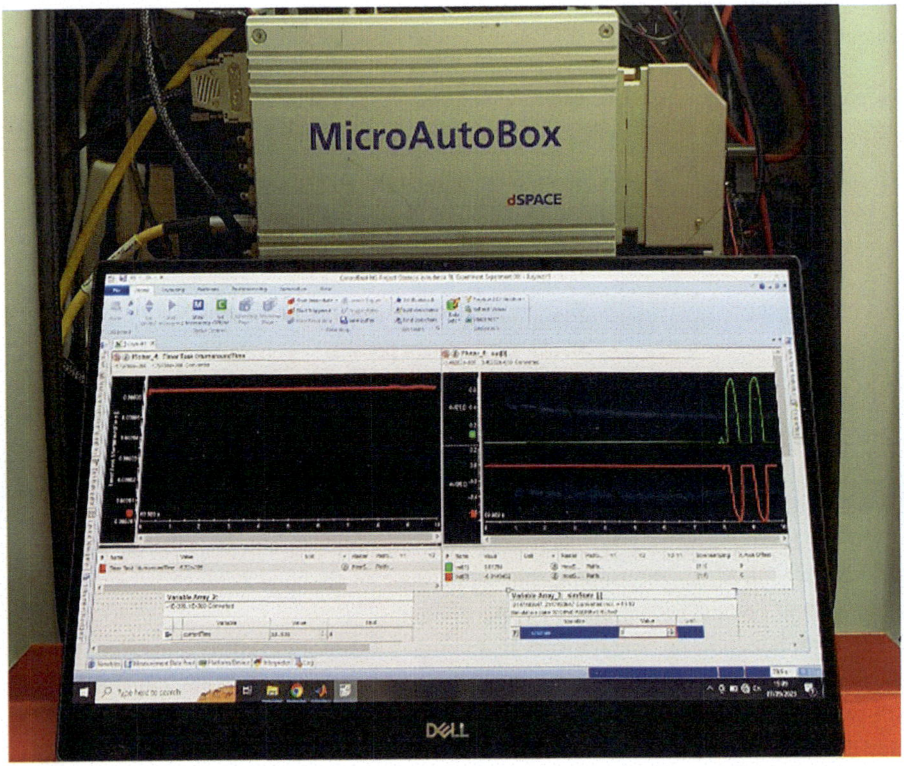

Fig. 5 Real-time implementation of the resulting agent on a dSPACE MicroAutoBox II system

4 Test Scenario and Simulation Results

4.1 Selected Maneuver

The schematic of the test track is shown in Fig. 6, and consists of three sectors, defined by cones. The reference path is the middle line of the course, which is filtered (with a 15 m moving average filter) according to the methodology in [2]. In the simulations, the vehicle is firstly accelerated in straight line to the target speed, before reaching sector 1 of the course, point at which the torque demand is set to zero. A single training episode includes both the initial acceleration and the cornering phases through sectors 1–3. The test is considered successful when the vehicle completes the maneuver without hitting any cone.

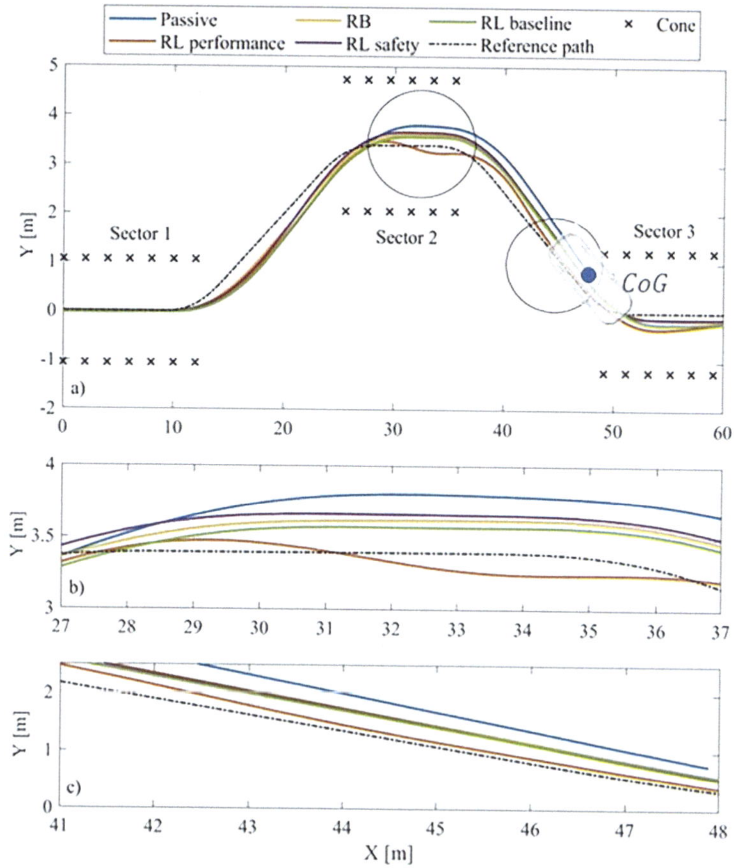

Fig. 6 Reference path for the obstacle avoidance maneuver, constraints defined by the cones, and simulation results in terms of vehicle path for the considered control configurations, including zoomed views of the most relevant sections of the test

4.2 Key Performance Indicators

The performance of the RL and RB TV strategies is assessed through the following KPIs:

- The vehicle exit speed, V_{fin}, i.e., for $X = 60$ m, which is associated with the overall level of vehicle agility.
- The maximum magnitude of the rear axle sideslip angle, $|\alpha_{R,max}|$, which is the main stability indicator.
- The maximum magnitude of the yaw rate, $|\dot{\psi}_{max}|$, since significant yaw rate peaks are usually symptomatic of very agile and potentially oversteering or unstable behavior.

- The root means square values of the yaw rate error $\Delta\dot{\psi}_{TV}$, rear axle sideslip angle constraint violation $\Delta\alpha_R\left(\Delta\alpha_R = \text{sat}_0^{\infty}\left(|\alpha_R| - \alpha_{R,th}\right)\right)$, and lateral and heading angle errors, e_y and $\Delta\psi$, w.r.t. the reference path:

$$RMS(e) = \sqrt{\frac{1}{t_f - t_i}\int_{t_i}^{t_f} e^2 dt} \tag{10}$$

- The normalized integrals of the absolute value of the steering wheel angle δ_{sw}, steering wheel rate $\dot{\delta}_{sw}$, reference direct yaw moment M_z (resulting from the torque corrections), and torque demand reduction ΔT, which describe the intensity and abruptness of the required steering action and TV controller interventions:

$$IA(c) = \sqrt{\frac{1}{t_f - t_i}\int_{t_i}^{t_f} |c| dt} \tag{11}$$

In (10) and (11), e and c are auxiliary variables, such that $e = \Delta\dot{\psi}_{TV}, \Delta\alpha_R, e_y, \Delta\psi$, and $c = \delta_{sw}, \dot{\delta}_{sw}, M_z, \Delta T$; and t_i and t_f are the initial and final times of the relevant part of the test, corresponding to the coordinates $X = 0$ m and $X = 60$ m in Fig. 6.

4.3 Simulation Results

Figures 6, 7 and 8 report the simulation results for the obstacle avoidance scenario, from an initial vehicle speed of 58 km/h, while Table 3 shows the respective KPIs. In these conditions, the passive configuration, i.e., the vehicle without TV control but still including the driving automation system, does not complete the test, and hits a cone at the beginning of sector 3, see the top view of the vehicle in Fig. 6. On the contrary, all proposed TV strategies enable the vehicle to successfully complete the maneuver, each of them with its own specificities.

The RB controller achieves the highest exit speed, with $V_{fin} = 45.9$ km/h, and the best yaw rate tracking performance, with $RMS\left(\Delta\dot{\psi}_{TV}\right) = 6.3°/\text{s}$, which also corresponds to the lowest $RMS(\Delta\psi)$ value. The smooth torque correction profile generates corrective direct yaw moments that tend to alternatively enhance agility (e.g., see the peak at $X =$ 10 m in Fig. 7a) and stability (e.g., see the peak at $X = 12$ m in the same subplot).

At the end of sector 1, differently from the RB strategy, the RL baseline agent directly applies a stabilizing direct yaw moment, which facilitates the following phases. In the

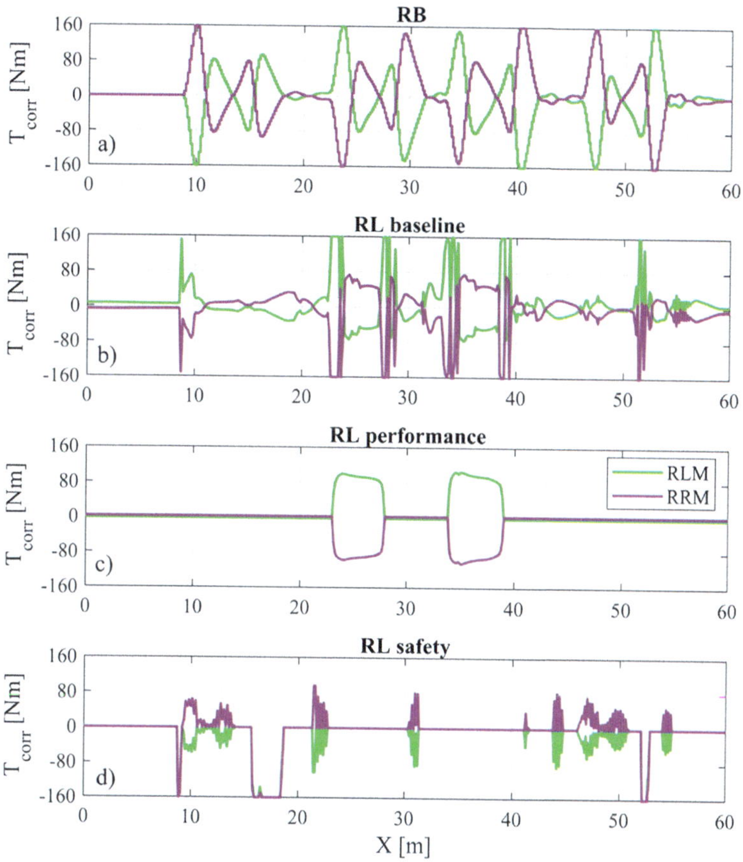

Fig. 7 Simulation results: control input profiles for the considered control configurations

most critical part of the test, ranging from the approach to sector 2 to the sector 3 entry, the signs of the T_{corr} peaks tend to be the same for the two configurations, see Fig. 7a, b. W.r.t. the RB controller, the baseline agent very marginally reduces $|e_y|$ and $|\Delta\psi|$ in the transition between sectors 2 and 3 (see the zoomed views in Fig. 6), where the passive vehicle fails.

The variation of the reward function scaling factors radically modifies the agent behavior. For example, for the RL performance agent, the reduction of R_{α_R} and $R_{\Delta\dot{\psi}}$ induces only two activations, during the transitions between sectors, see Fig. 7c. Both TV interventions correspond to destabilizing direct yaw moments, which bring the car to controlled oversteering and drifting behaviors, with α_R peaks exceeding 12°, see Fig. 8d. The benefit is an evident reduction of $|e_y|$ during sector 2, as well as in the transition from sector 2 to sector 3, see the zoomed subplots of Fig. 6. This results in the lowest $RMS(e_y)$ value, i.e., 0.23 m, among the considered configurations, see Table 3. Such trajectory tracking

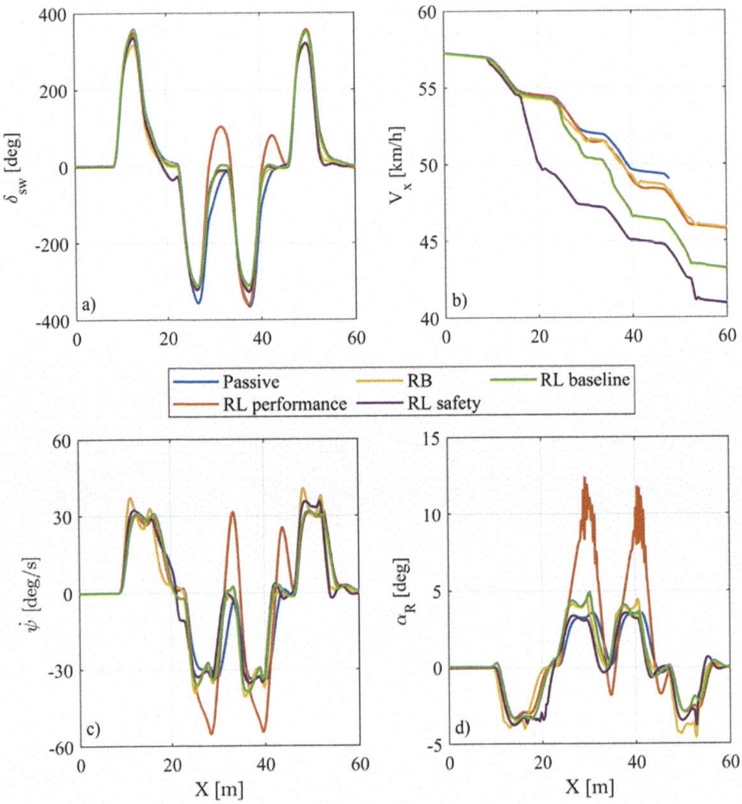

Fig. 8 Simulation results: steering input profiles and main vehicle response variables

improvement is caused by the influence of the cone-related boundaries expressed through the R_{cone} contribution of the reward function used in the training phase, even if the resulting NN of the online algorithm does not receive any path- or cone-related input, see (2). Also, IA_{M_z} is smaller than for the RB and RL baseline implementations, and the exit speed is comparable to the one of RB. However, the vehicle response is more difficult to handle for the driver, see Fig. 8a and the steering effort indicators in Table 3, which are the highest among the controllers. Nevertheless, the philosophy—enforced by the RL performance algorithm—of pushing the boundaries of the current generation of stability controllers through automated driving behaviors adopting expert driving techniques is a recent and promising research trend, which could prevent road accidents in emergency conditions [17].

On the contrary, for the RL safety agent, a lower penalization on $R_{\Delta T}$ induces the algorithm to slow down the vehicle in the transition between sectors 1 and 2. Interestingly, this behavior is similar to the one of the pre-emptive braking function in [1]. The lower

Table 3 Performance indicators for the considered control strategies during the obstacle avoidance maneuver: (i) the rows corresponding to the most relevant indicators are highlighted in grey; (ii) the KPIs for the best performing controllers are reported in bold; and (iii) the KPIs for the best performing RL configuration are underlined in bold

KPI	RB	RL baseline	RL perf.	RL safety
V_{fin} [km/h]	**45.9**	43.2	**<u>45.8</u>**	40.9
$\lvert\alpha_{R,max}\rvert$ [°]	4.56	4.91	12.38	**<u>3.78</u>**
$\lvert\dot{\psi}_{max}\rvert$ [°/s]	41.0	38.7	55.0	**<u>35.9</u>**
$RMS(\Delta\alpha_R)$ [°]	0	0	1.60	0
$RMS(\Delta\dot{\psi}_{TV})$ [°/s]	**6.3**	7.5	16.0	**<u>7.3</u>**
$RMS(e_y)$ [m]	**0.25**	0.28	**<u>0.23</u>**	0.25
$RMS(\Delta\psi)$ [°]	**2.3**	2.7	3.0	**<u>2.4</u>**
$IA(\delta_{sw})$ [°]	103	109	118	104
$IA(\dot{\delta}_{sw})$ [°/s]	593	616	741	**<u>578</u>**
$IA(M_z)$ [N m]	1087	814	468	167
$IA(\Delta T)$ [N m]	0	0.1	0.4	19.5

speed inherently reduces the following peaks and oscillations on α_R and $\Delta\dot{\psi}_{TV}$ w.r.t. the other algorithms. Hence, the vehicle easily passes the test with low values of the steering effort KPIs, e.g., $IA(\dot{\delta}_{sw})$ is the lowest among the TV controllers. Although this solution was achieved with an automated path tracking controller, such TV behavior would be especially suitable for a human driver in everyday road scenarios. The braking action of the RL safety strategy allows the vehicle to pass the test also if the initial speed is increased up to 60 km/h, while the other RL-based algorithms and the RB strategy fail.

5 Conclusion

This study evaluated real-time capable torque-vectoring (TV) algorithms based on reinforcement learning (RL). An RL agent was designed for the obstacle avoidance scenario, and was compared with a baseline rule-based (RB) TV controller, through simulations with an experimentally validated model of an automated electric vehicle with multiple powertrains.

With the same general formulation in terms of state space, reward function structure, and network size, the RL agent can provide significantly different solutions, according to the scaling factors of the reward function that guides the training. In the specific analysis, the agent was trained as: (i) a baseline TV system providing balanced performance w.r.t. the different requirements, similarly to the RB controller; (ii) a more performance-oriented set-up, for which controlled drifting enhances path tracking and exit speed; and (iii) a safety-oriented controller, prioritizing ease of maneuvering through pre-emptive speed reductions.

The simulation results highlighted the radical diversity of behaviors of the considered RL agent declinations, in terms of control inputs and vehicle response, which would not be possible through the tuning of the baseline feedback TV controller. Nevertheless, in comparison with the RL agents, the RB strategy provides a very effective and balanced response, with excellent yaw rate tracking and the highest exit speed, even if, for example, the RL safety algorithm can complete the test from higher entry speeds.

The diversity of behaviours was possible thanks to: (i) the state space definition, which provides a broad overview of the vehicle's lateral dynamics; (ii) the choice and tuning of the reward function, which clearly describes the target the agent must achieve during training, avoiding chattering and excessive control input; and (iii) the high-fidelity simulation environment, which includes powertrain dynamics, actuation delays, and nonlinear tire behaviour.

The ability of the proposed RL agent to learn such a control task in a high-fidelity simulation environment and complex scenario can potentially bring multiple benefits in the near future, such as: (i) improving the design of more advanced control strategies involving multiple actuators and different targets, through appropriate definitions of the reward function, state space, and environment; or (ii) supporting the development of benchmarking control strategies by demonstrating the optimal solution for the specific scenario. As a further point, although the computational effort required for the training process is relevant, the trained controller is computationally efficient.

Future work will include: (a) further development of the RL agent architecture and training process to achieve a clear performance benefit w.r.t. the RB TV algorithm; (b) the generalization of the RL agents to a multitude of scenarios and tire-road friction conditions; (c) the analysis of the benefits of including the road scenario information (e.g., the lane boundaries) in the training phase of the TV agents, w.r.t. training based on more conventional performance indicators; and (d) comparisons with more advanced model-based controllers, such as model predictive TV controllers.

References

1. Guastadisegni, G., So, K.M., Parra, A., Tavernini, D., Montanaro, U., Gruber, P., Soria, L., Mantriota, G., Sorniotti, A.: Vehicle stability control through pre-emptive braking. Int. J. Automot. Technol. **24**(2), 347–365 (2023)
2. Chatzikomis, C., Sorniotti, A., Gruber, P., Zanchetta, M., Willans, D., Balcombe, B.: Comparison of path tracking and torque-vectoring controllers for autonomous electric vehicles. IEEE Trans. Intell. Veh. **3**(4), 559–570 (2018)
3. Lenzo, B., Zanchetta, M., Sorniotti, A., Gruber, P., De Nijs, W.: Yaw rate and sideslip angle control through single input single output direct yaw moment control. IEEE Trans. Control Syst. Technol. **29**(1), 124–139 (2021)

4. Mazzilli, V., De Pinto, S., Pascali, L., Contrino, M., Bottiglione, F., Mantriota, G., Gruber, P., Sorniotti, A.: Integrated chassis control: classification, analysis and future trends. Annu. Rev. Control **51**, 172–205 (2021)
5. De Novellis, L., Sorniotti, A., Gruber, P., Shead, L., Ivanov, V., Hoepping, K.: Torque vectoring for electric vehicles with individually controlled motors: state-of-the-art and future developments. In: 26th Electric Vehicle Symposium (2012)
6. Asperti, M., Vignati, M., Sabbioni, E.: On torque vectoring control: review and comparison of state-of-the-art approaches. Machines **12**(3) (2024)
7. Wei, H., Zhang, N., Liang, J., Ai, Q., Zhao, W., Huang, T., Zhang, Y.: Deep reinforcement learning based direct torque control strategy for distributed drive electric vehicles considering active safety and energy saving performance. Energy **238**(part B) (2022)
8. Coppola, A., Petrillo, A., Rizzo, R., Santini, S.: Adaptive cruise control for autonomous electric vehicles based on Q-learning algorithm. In: AEIT International Annual Conference (2021)
9. Kuutti, S., Bowden, R., Jin, Y., Barber, P., Fallah, S.: A survey of deep learning applications to autonomous vehicle control. IEEE Trans. Intell. Transp. Syst. **22**(2), 712–733 (2021)
10. Yurtsever, E., Lambert, J., Carballo, A., Takeda, K.: A survey of autonomous driving: common practices and emerging technologies. IEEE Access **8**, 58443–58469 (2020)
11. Tampuu, A., Matiisen, T., Semikin, M., Fishman, D., Muhammad, N.: A survey of end-to-end driving: architectures and training methods. IEEE Trans. Neural Netw. Learn. Syst. **33**(4), 1364–1384 (2022)
12. Morales, M.: Grokking Deep Reinforcement Learning. Manning Publications (2020)
13. Sutton, R., Barto, A.G.: Reinforcement Learning: An Introduction. MIT Press, Cambridge (2018)
14. Hoffmann, P., Gorelik, K., Ivanov, V., Bosch, R.: Comparison of reinforcement learning and model predictive control for over-actuated systems. In: 15th International Symposium on Advanced Vehicle Control (2022)
15. AVL VSM™ | AVL, by AVL List GmbH, Hans-List-Platz 1, 8020, Graz (AT), last consulted on 26th March 2025
16. MATLAB and Reinforcement Learning Toolbox Release 2023a. The MathWorks, Inc., Natick, Massachusetts, United States
17. Stano, P., Montanaro, U., Tavernini, D., Tufo, M., Fiengo, G., Novella, L., Sorniotti, A.: Model predictive path tracking control for automated road vehicles: a review. Annu. Rev. Control **55**, 194–236 (2023)

Performance Evaluation of Path Tracking Controllers for Scaled Robotic Cars

Carmine Caponio, Pietro Stano, Giulio Scattolini, Nuccio Lo Bello, Michael Stickley, Mario Mihalkov, Alessandro Vigliani, Alessandro Rizzo, Aldo Sorniotti, and Umberto Montanaro

Abstract

A versatile platform for testing automated driving algorithms is represented by scaled robotic cars, which can reduce costs, increase the repeatability of testing, and enhance safety, compared to experiments conducted on full-size vehicles. Therefore, scaled cars are used before testing control solutions on real vehicle demonstrators. Similarly to full-size autonomous vehicles, scaled robotic cars typically feature a hierarchical control structure that encompasses various functionalities, including path planning and path tracking, as well as state estimation algorithms. This work presents an experimental comparison of path tracking control methods applied to scaled robotic vehicles with front steering actuation. Six path tracking control strategies—namely, based on (i) model predictive control, (ii) pole placement control, (iii) linear quadratic regulation, (iv) sliding mode control, (v) proportional integral control, and (vi) the Stanley method—are designed, implemented, and experimentally evaluated on scaled robotic QCar vehicles along five paths at eight vehicle speeds. Performance is assessed through a set of key performance indicators, demonstrating the superiority of the model predictive and sliding mode controllers.

C. Caponio · P. Stano · M. Stickley · M. Mihalkov · U. Montanaro (✉)
School of Engineering, University of Surrey, Guildford, UK
e-mail: u.montanaro@surrey.ac.uk

G. Scattolini · N. Lo Bello · A. Vigliani · A. Sorniotti
Department of Mechanical and Aerospace Engineering, Politecnico di Torino, Turin, Italy

A. Rizzo
Department of Electronics and Telecommunications, Politecnico di Torino, Turin, Italy

D. Watzenig (ed.), *Recent Advances in Autonomous Vehicle Technology–Perception and Path Planning*, Sustainable Mobility & Energy,
https://doi.org/10.1007/978-3-032-01001-8_6

Keywords

Scaled robotic cars • Path tracking control • Model predictive control • Sliding mode control • Pole placement • Linear quadratic regulator • Stanley controller • Experimental comparison

1 Introduction

With the widespread advancement of Industry 4.0 globally, mobile robotic systems have garnered considerable attention within the industry and academia [1], and are used in various applications in laboratories, industrial settings, warehousing, and transportation, just to name a few [2]. In this framework, scaled robotic cars are commonly adopted as testing platforms for automated driving algorithms [3, 4], including path planning [5, 6], path tracking (PT) [7], and vehicle platooning [8, 9]. This is due to their ability to offer: (i) cost-effective testing, leading to cost savings compared to full-scale vehicle experiments; (ii) improved repeatability through better control of the testing environment, thus enhancing fairness when comparing different control algorithms; and (iii) increased safety, as scaled robotic cars pose lower risks in the event of system failures or collisions.

Within automated vehicle applications, PT controllers play a key role in the control architecture, as they define the actuator commands—typically expressed in terms of steering angle and traction/braking torque levels—to follow the paths determined by higher level controllers, e.g., the path planners, while adapting to dynamic and uncertain real-world conditions [10–12]. In this context, several previous studies, e.g., see [13, 14], delve into assessing the influence of design factors, such as the vehicle model complexity and the robustness of the control scheme, across various PT control strategies. The considered algorithms include geometry-based implementations (pure pursuit and Stanley methods), robust controllers (sliding mode controllers (SMCs) and H_∞ strategies), and optimal controllers (linear quadratic regulators (LQRs) and model predictive controllers (MPCs)).

To complement simulations, scaled autonomous vehicles are used in the literature to experimentally assess the effectiveness of PT control solutions [15–19]. For instance, reference [15] introduces a hierarchical control strategy consisting of: (i) a feedforward-feedback control structure, determining the desired yaw rate based on the reference curvature and a variable look-ahead distance; and (ii) a model reference adaptive controller based on a kinematic reference model, generating the steering angle to track the reference yaw rate. In [16], an active disturbance rejection control system is introduced for lane keeping applications. This strategy, based on a single-track model (STM), has experimentally demonstrated superior tracking performance compared to a proportional integral derivative (PID) controller. In [17], a robust linear matrix inequality based MPC is designed to track the reference path while reducing the progress of battery wear. The studies in [18, 19] include other examples of MPC PT solutions.

Despite the wealth of PT solutions in the literature, only few papers provide experimental comparisons among benchmarking PT controllers. For example, reference [20] compares state-of-the-art model-free and model-based control strategies on a full-scale electric vehicle. In this subject area, this work presents the design and the experimental assessment of PT strategies for scaled robotic cars, i.e., QCars [21]. Specifically, the contributions are:

- The formulation of six benchmarking PT controllers for automated scaled vehicles, i.e., MPC, Pole placement (PP), LQR, SMC, Proportional integral (PI) and Stanley (STC) PT strategies. With the exception of PI and STC, the design is based on the experimentally validated STM proposed in [22];
- The experimental assessment of the proposed strategies through five pre-defined paths at eight vehicle speed levels, and a set of key performance indicators (KPIs).

The PT control solutions are implemented within a control architecture that also includes a longitudinal speed controller and state estimators. The use of the same control architecture and operating vehicle conditions ensures the objective and fair comparison of the PT control methods.

The remainder is organised as follows: Sect. 2 provides a brief overview of the specific scaled robotic car system setup; Sect. 3 details the vehicle model; the control architecture hosting the PT controllers is outlined in Sect. 4; Sect. 5 presents and discusses the experimental results; finally, conclusions are drawn in Sect. 6.

2 Experimental Setup and Simulation Framework

The robotic scaled car used to validate the PT control strategies is the self-driving QCar platform by Quanser, depicted in Fig. 1 [21]. This open-architecture system is specifically designed for academic research. Configured as all-wheel drive (AWD), the robotic car is powered by a direct current (DC) electric motor, with its rotational speed measured by an encoder, and operates with a single-speed gearbox and appropriate mechanical differentials. The steering actuation is applied by a servo motor, which ensures a maximum steering angle of approximately $\pm 30^{\circ}$. Additionally, the QCar is equipped with an onboard 9-axis inertial measurement unit (IMU), and a lidar (laser imaging detection and ranging) sensor for environmental perception. The onboard computing system includes an NVIDIA Jetson TX2 processor, incorporating a Quad-Core ARM Cortex-A57 microcontroller. Communication between the QCar and the ground station in Fig. 2 is established through WiFi. The proposed algorithms are developed in MATLAB 2022b/Simulink [23].

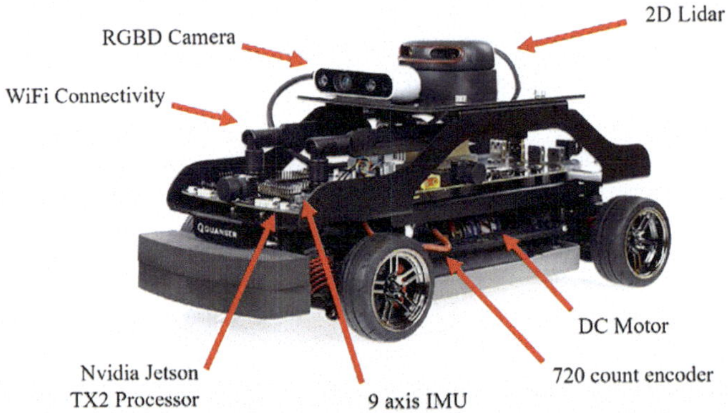

Fig. 1 The Quanser QCar with its main features [21]

Fig. 2 Base station and QCar available at the surrey team for the control of smart multi-agent systems operating autonomously and synergistically (Su-COSMOS)

3 Experimentally Validated Vehicle Model

This section introduces the vehicle model, based on a dynamic single-track formulation, used for the design of most of the proposed PT control strategies, encompassing the longitudinal and lateral vehicle dynamics, which are briefly discussed in Sects. 3.1 and 3.2. Figure 3 is a top view schematic of the robotic car along with the key variables and respective sign conventions. The reader is referred to [22] for the details on the model identification and experimental validation.

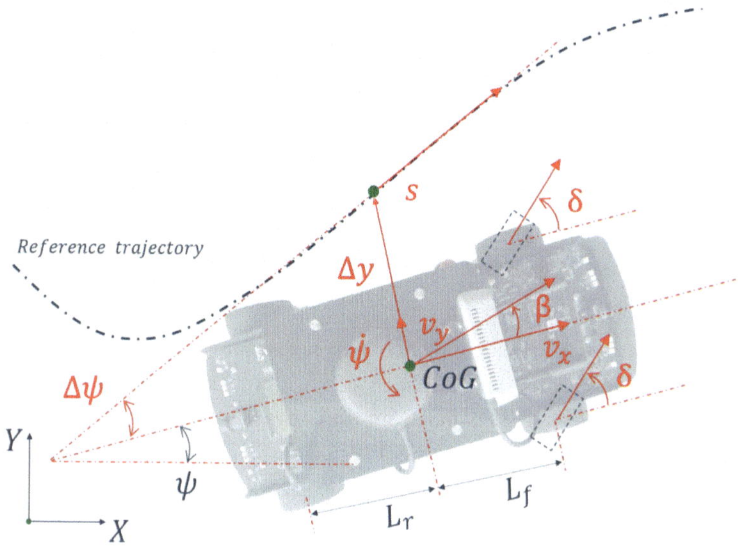

Fig. 3 Reference path and relative position and heading errors

3.1 Longitudinal Dynamics

Under the assumptions of forward motion of the scaled robotic vehicle, negligible longitudinal tyre slip, small sideslip angle magnitude, and negligible electric dynamics of the DC motor, the longitudinal vehicle dynamics are described by:

$$\begin{aligned} \dot{\omega} &= P_1 V_a - P_2 \omega - P_3 \text{sign}(\omega) \\ v &= \frac{R_w}{\mathcal{G}} \omega \end{aligned} \tag{1}$$

where v is the vehicle speed; ω is the angular speed of the motor rotor; V_a is the armature voltage applied to the DC motor; $\mathcal{G}$ is the transmission ratio; R_w is the wheel radius; and $P_1 = K_t/(J_w R_a)$, $P_2 = (K_t K_v/R_a + B)/J_w$, and $P_3 = T_c/J_w$ are lumped parameters, in which J_w is the equivalent mass moment of inertia of the whole system, R_a is the armature resistance, B is an equivalent viscous friction coefficient, K_t is the torque constant, K_v is the motor back electromotive force constant, and T_C is the Coulomb friction torque.

3.2 Lateral Dynamics

By considering a dynamic STM and under the assumption of small sideslip angles, the lateral velocity and yaw rate dynamics are described by:

$$
\begin{aligned}
mv\dot{\beta} &= -(C_f + C_r + m\dot{v})\beta - \left(\frac{C_f L_f}{v} - \frac{C_r L_r}{v} + mv\right)r + C_f\delta \\
I_G\dot{r} &= -(C_f L_f - C_r L_r)\beta - \left(\frac{C_f L_f^2}{v} + \frac{C_r L_r^2}{v}\right)r + C_f L_f \delta
\end{aligned}
\tag{2}
$$

where β is the vehicle sideslip angle; δ is the steering angle; and $r = \dot{\psi}$ is the yaw rate, with ψ being the heading angle. The system parameters are the yaw mass moment of inertia I_G, the vehicle mass m, the front and rear semi-wheelbases L_f and L_r, and the front and rear axle cornering stiffness C_f and C_r. While L_f and L_r are known and provided by the robotic vehicle manufacturer, C_f and C_r were identified through the two-stage least-squares method in [22].

The coordinates of the QCar centre of gravity (CG) in the inertial reference frame, denoted as X and Y, are obtained through the integration of:

$$
\begin{aligned}
\dot{X} &= v\cos(\beta + \psi) \\
\dot{Y} &= v\sin(\beta + \psi).
\end{aligned}
\tag{3}
$$

3.3 Experimental Validation

Examples of experimental validation results of the longitudinal and lateral dynamics model in Sects. 3.1 and 3.2 are shown in Fig. 4, for step inputs in terms of reference speed and steering angle. The accurate match between simulations and experiments makes the reference model a reliable tool for control system design.

4 Control Architecture

4.1 Control System Hierarchy

The experimental and control architecture is shown in Fig. 5, and includes:

- The Federated Extended Kalman Filter (FEKF), which collects the signals from the sensors installed on the vehicle, and estimates the states required by the control functions. The system adopts the FEKF formulation proposed in [22], consisting of two

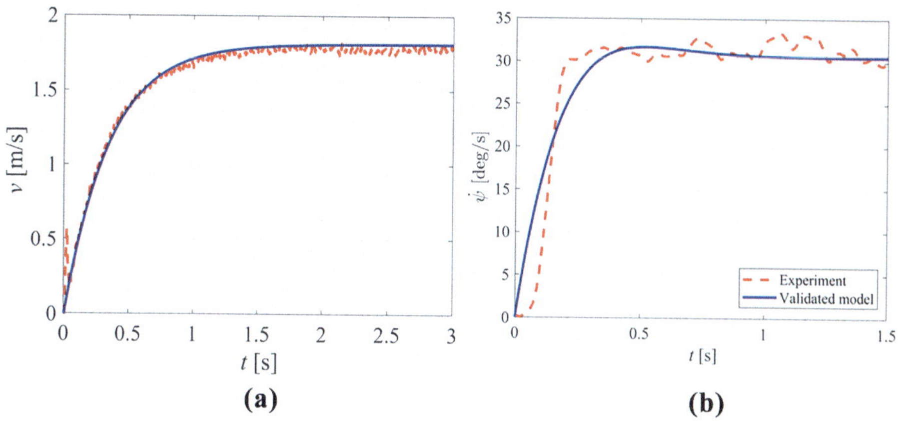

Fig. 4 Vehicle model validation. Experimental measurements (red dashed lines) and vehicle model results (blue solid lines) in terms of time (t) profiles of: **a** longitudinal speed for a step input profile of the reference speed; and **b** yaw rate for a step steering input

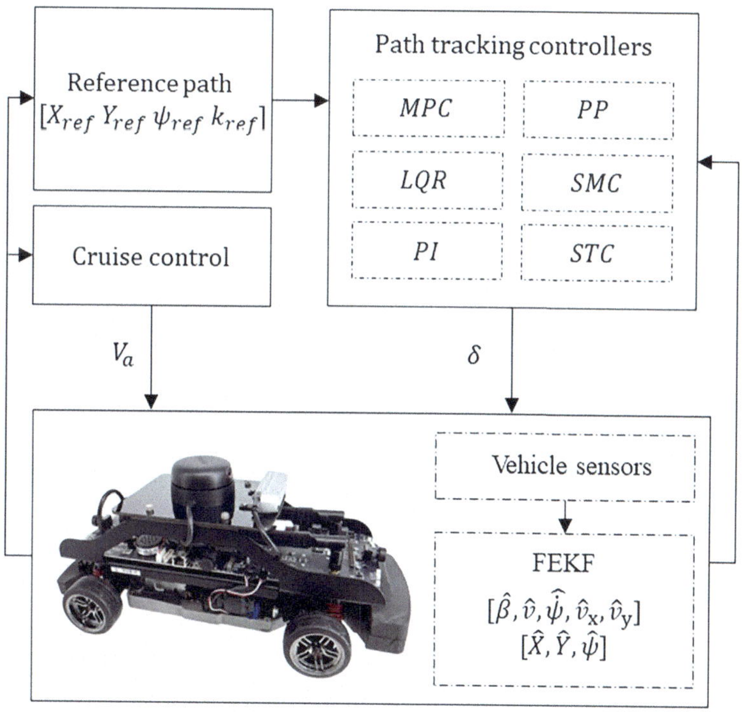

Fig. 5 Simplified schematic of the control architecture

local EKFs and a main filter. This configuration enhances the accuracy of the positioning system, by combining the position estimates from the local EKFs and by effectively using all the available onboard sensors, including the IMU, Lidar, and encoder. In Fig. 5, the additional notation '^' on top of the symbol of the variable denotes the estimate of the corresponding vehicle state.

- The path generator, which defines the coordinates of the reference path, X_{ref} and Y_{ref}, the reference heading angle of the vehicle, ψ_{ref}, and the reference curvature, k_{ref}.
- The PT controllers, computing the required steering angle to track the reference path.
- The PI-based cruise controller, which determines the input voltage V_a for the DC motor, to achieve the desired speed throughout the manoeuvres.

4.2 Reference Path and Vehicle Coordinates

The path generator dynamically plans the reference path, considering the navigation objectives and potential obstacles. As this work focuses on the PT function and the experimental comparison among different PT solutions, the reference profiles remain constant throughout the manoeuvre, to prevent coupling with the path planning layer, which could otherwise affect the objectivity of the PT algorithm assessment. In this framework, the reference path coordinates, X_{ref} and Y_{ref}, are expressed in the global coordinate system. The reference heading angle (ψ_{ref}) and the reference curvature (k_{ref}) are derived from the reference coordinates, and serve as inputs for the path tracking controllers. The heading angle error, $\Delta\psi$, and the lateral position error, Δy, of the vehicle's CG in relation to the reference path are defined as illustrated in Fig. 3. The mathematical formulation for Δy and $\Delta\psi$ are detailed in [24], and they depend on the longitudinal and lateral speeds, v_x and v_y, as well as the coordinates of the vehicle's CG, X and Y.

4.3 Overview of the Implemented Path Tracking Controllers

This section briefly describes the control formulations of the PT solutions that have been designed and implemented in the control architecture in Fig. 5.

4.3.1 Model Predictive Control

The MPC algorithm is based on a state-space formulation of the plant, also referred to as the prediction or internal model, whose dynamics are described as follows:

$$\dot{x}_{MPC} = f(x_{MPC}, P_{MPC}, u_{MPC}) \tag{4}$$

where x_{MPC} is the state vector; P_{MPC} is the system parameter vector; and u_{MPC} is the control input. At each sampling time, the target of the MPC strategy is to minimize a cost function J_{MPC}, based on the predicted system dynamics (4) along the horizon H_p.

For the proposed MPC strategy, the internal model state vector consists of the heading angle error, the lateral error, and their time derivatives (i.e., $x = \left[\Delta y \Delta \dot{y} \Delta \psi \Delta \dot{\psi}\right]$), augmented with the steering angle, thus resulting into $x_{MPC} = \left[x\ \delta\right]^T$. The parameter vector is selected as $P_{MPC} = v_x$, while the control input is the steering angle derivative, i.e., $u_{MPC} = \dot{\delta}$. Selecting $\dot{\delta}$ as control input allows for weighing the rate of change of the steering angle in the MPC optimisation problem, which limits steering angle oscillations and chattering.

Based on the linear lateral vehicle dynamics formulation in (2), the error dynamics are given by [25]:

$$\dot{x} = Ax + B_1\delta + B_2 k_{ref} \tag{5}$$

where A, B_1, and B_2 are:

$$A = \begin{bmatrix} 0 & 1 & 0 & 0 \\ 0 & -\frac{C_f + C_r}{m v_x} & \frac{C_f + C_r}{m} & \frac{C_f L_f + C_r L_r}{m v_x} \\ 0 & 0 & 0 & 1 \\ 0 & -\frac{L_f C_f - L_r C_r}{v_x I_z} & \frac{C_f L_f - C_r L_r}{I_z} & -\frac{L_f^2 C_f + L_r^2 C_r}{v_x I_z} \end{bmatrix}$$
$$B_1^T = \begin{bmatrix} 0 & \frac{C_f}{m} & 0 & \frac{L_f C_f}{I_z} \end{bmatrix}$$
$$B_2^T = \begin{bmatrix} 0 & -\frac{C_f L_f - C_r L_r}{m} - v_x^2 & 0 & -\frac{L_f^2 C_f + L_r^2 C_r}{v_x I_z} \end{bmatrix} \tag{6}$$

Consequently, the prediction model is:

$$\dot{x}_{MPC} = A_a x_{MPC} + B_{1,a} u_{MPC} + B_{2,a} k_{ref} \tag{7}$$

where the augmented matrices A_a, $B_{1,a}$ and $B_{2,a}$ are:

$$A_a = \begin{bmatrix} A & B_1 \\ \mathcal{O}_{1,4} & 0 \end{bmatrix} \in \mathbb{R}^{5\times 5}\ B_{1,a} = \begin{bmatrix} \mathcal{O}_{4,1} \\ 1 \end{bmatrix} \in \mathbb{R}^{5\times 1}\ B_{2,a} = \begin{bmatrix} B_2 \\ 0 \end{bmatrix} \in \mathbb{R}^{5\times 1} \tag{8}$$

with $\mathcal{O}_{n,m}$ being the null element of the vector space $\mathbb{R}^{n\times m}$.

The MPC control law minimises J_{MPC}, subject to a set of equality and inequality constraints [26]. For the proposed MCP-PT problem, J_{MPC} is quadratic and defined as:

$$J_{MPC} = \int_0^{H_p} l(x_{MPC}(\tau), u_{MPC}(\tau))d\tau + ||\Delta y(H_p)||^2_{Q_{MPC,f}}$$

with

$$l(x_{MPC}(\tau), u_{MPC}(\tau)) = ||x_{MPC}(\tau)||^2_{Q_{MPC}} + ||u_{MPC}(\tau)||^2_{R_{MPC}} \tag{9}$$

where $\int_0^{H_p} l(x_{MPC}(\tau), u_{MPC}(\tau))d\tau$ is the time integral of the weighted and squared 2-norm of the predicted state tracking error and control input along H_p, while $\Delta y(H_p)^2_{Q_{MPC,f}}$ is the terminal cost. The weight matrices, $Q_{MPC} \in \mathbb{R}^{5\times5}$, $Q_{MPC,f} \in \mathbb{R}^{1\times1}$, and $R_{MPC} \in \mathbb{R}^{1\times1}$, are positive matrices, and are scheduled based on the longitudinal speed of the trajectory tracking test. The optimal control problem is then formulated as:

$$\min_{\text{u}} J_{MPC}(x_{MPC}, P_{MPC}, u_{MPC})$$

s.t.

$$\begin{aligned} \dot{x}_{MPC} &= A_a x_{MPC} + B_{1,a} u_{MPC} + B_{2,a} k_{ref} \\ x_{MPC}(0) &= x_{MPC,t} \\ |\delta| &\leq \delta_{max} \end{aligned} \tag{10}$$

where $x_{MPC,t}$ is the initial value of the state of the robotic car for the optimisation problem (i.e., the vehicle state at the current time instant t), and δ_{max} is the maximum steering angle, accounting for the mechanical saturation of the vehicle's steering actuation system. According to the receding horizon approach, only the first sample of the optimal control sequence is applied to the robotic car, and the solution of the optimal control problem (10) is repeated at the next sampling time with an updated value for $x_{MPC,t}$.

The controller sampling time T_s was chosen based on the FEKF sampling time of 10 ms, while the prediction horizon H_p was selected through a simulation-based sensitivity analysis, targeting a satisfactory trade-off between closed-loop tracking performance and computational time. This resulted in an optimal H_p value of 100 ms. The MPC algorithm was implemented through the ACADO toolkit [27], with the following settings: Gaussian Netwon Hessian approximation, multiple shooting discretization, fourth order implicit Runge Kutta integrator, and qpOASES solver.

4.3.2 Pole Placement

The proposed pole placement (PP) strategy is based on the approach presented in [25], where the tracking error dynamics (5) are stabilised and controlled via a linear full-state feedback and feedforward control action, i.e.:

$$\delta = -K_{PP}x + K_{ff,PP}k_{ref} \tag{11}$$

where K_{PP} and $K_{ff,PP}$ are the feedback and feedforward gains. By using the control action in (11), the closed-loop dynamics are:

$$\begin{aligned} \dot{x}_{PP} &= (A - B_1K_{PP})x_{PP} + \big(B_1K_{ff,PP} + B_2\big)k_{ref} \\ &= A_{cl}x_{PP} + B_{cl}k_{ref} \end{aligned} \tag{12}$$

where $x_{PP} = x$ is the closed-loop state vector.

According to [25], the feedforward gain $K_{ff,PP}$ is designed to have zero steady-state lateral error in presence of constant curvatures, thus:

$$K_{ff,PP} = \frac{mv_x^2}{L}\left(\frac{L_r}{C_f} - \frac{L_f}{C_r} + \frac{L_f}{C_r}K_{PP,3}\right) + L - L_rK_{PP,3} \tag{13}$$

with $K_{PP,3}$ being the third component of K_{PP} and $L = L_r + L_f$.

The dynamics of the controlled system (12) depend on the eigenvalues of the closed-loop matrix A_{cl}. In the proposed PP strategy, the desired set of eigenvalues λ_{des} to impose to A_{cl} through the feedback gain K_{PP}, e.g., via Ackermann's formula, is selected by optimising the closed-loop transfer functions. Specifically, the closed-loop transfer matrix is:

$$G(s) = (s_{\mathcal{L}}I_4 - A_{cl})^{-1}B_{cl}$$

with

$$[G_1(s_{\mathcal{L}})G_2(s_{\mathcal{L}})G_3(s_{\mathcal{L}})G_4(s_{\mathcal{L}})]^T = \left[\frac{\Delta y(s_{\mathcal{L}})}{k_{ref}(s_{\mathcal{L}})}\frac{\Delta \dot{y}(s_{\mathcal{L}})}{k_{ref}(s_{\mathcal{L}})}\frac{\Delta \psi(s_{\mathcal{L}})}{k_{ref}(s_{\mathcal{L}})}\frac{\Delta \dot{\psi}(s_{\mathcal{L}})}{k_{ref}(s_{\mathcal{L}})}\right]^T \tag{14}$$

where $s_{\mathcal{L}}$ is the Laplace variable, and $I_4 \in \mathbb{R}^{4\times 4}$ is the identity matrix. The optimal set of eigenvalues λ^*_{des} is determined through the following static optimisation problem:

$$\lambda^*_{des} = \arg\min_{\lambda_{des}}\{J_{PP}(\lambda_{des})\}$$

$$J_{PP}(\lambda_{des}) = \alpha_1 ||G_1(s_{\mathcal{L}}; \lambda_{des})||_\infty^2 + \alpha_2 ||G_3(s_{\mathcal{L}}; \lambda_{des})||_\infty^2 + \alpha_3 |G_3(0; \lambda_{des})|^2$$

s.t.

$$\begin{aligned} &\mathcal{R}e\{\lambda_{des,i}\} < 0, i = 1, \ldots, 4 \\ &K_{PP} = e_4^T \mathcal{C}^{-1}(A, B_1)\gamma_c(A; \lambda_{des}) \\ &K_{ff,PP} = \frac{mv_x^2}{L}\left(\frac{L_r}{C_f} - \frac{L_f}{C_r} + \frac{L_f}{C_r}K_{PP,3}\right) + L - L_r K_{PP,3} \end{aligned} \tag{15}$$

where the constraint $\mathcal{R}e\{\lambda_{des,i}\} < 0$ is required for imposing closed-loop stability; $K_{ff,PP}$ is computed via (13); K_{PP} is derived through Ackermann's formula, thus $e_4^T = \begin{bmatrix} 0 & 0 & 0 & 1 \end{bmatrix}$; $\mathcal{C}(A, B_1)$ is the observability matrix of the pair (A, B_1); and, by defining $\gamma_c(s_{\mathcal{L}}; \lambda_{des})$ as the fourth order polynomial with roots λ_{des}, $\gamma_c(A; \lambda_{des}) \in \mathbb{R}^{4\times4}$ is the matrix obtained by replacing in γ_c the variable $s_{\mathcal{L}}$ with A. The cost function of the optimisation problem (15) is the weighted sum of: (i) the infinity norm of the transfer function for the lateral error (i.e., $G_1(s_{\mathcal{L}}; \lambda_{des})_\infty$); (ii) the infinity norm of the orientation error transfer function (i.e., $G_3(s_{\mathcal{L}}; \lambda_{des})_\infty$); and (iii) the static gain of the orientation error (i.e., $|G_3(0; \lambda_{des})|$). The weights α_1, α_2 and α_3 are used to prioritise the different terms. The optimisation problem (15) is solved numerically via the use of the MATLAB optimisation toolbox for different longitudinal vehicle speeds, and the resulting control gains are collected in look-up tables.

4.3.3 Linear Quadratic Regulator

The design of the LQR control strategy is based on the formulation of the path tracking error dynamics in [28]. Hence, by defining $x_{LQR} = \left[v_y \dot{\psi} \Delta y \Delta \psi\right]^T$ as the vehicle state vector, where v_y is the lateral vehicle speed, the system dynamics are given by:

$$\dot{x}_{LQR} = A_{path} x_{LQR} + B_{1,path}\delta + B_{2,path}k_{ref} \tag{16}$$

where the system matrices are:

$$A_{path} = \begin{bmatrix} \frac{C_f + C_r}{mv_x} & \frac{L_f C_f + L_r C_r}{mv_x} - v_x & 0 & 0 \\ \frac{L_f C_f - L_r C_r}{v_x I_z} & \frac{L_f^2 C_f + L_r^2 C_r}{v_x I_z} & 0 & 0 \\ -1 & 0 & 0 & v_x \\ 0 & -1 & 0 & 0 \end{bmatrix}$$

$$B_{1,path}^T = \left[-\frac{C_f}{m} \; -\frac{L_f C_f}{I_z} \; 0 \; 0 \right]$$
$$B_{2,path}^T = \left[0 \; 0 \; 0 \; v_x \right] \tag{17}$$

The control action for the LQR solution is given by:

$$\delta = -K_{LQR} x_{LQR} + K_{ff,LQR} k_{ref} \tag{18}$$

where $K_{ff,LQR}$ is a feedforward gain; and K_{LQR} is the feedback control gain computed by solving the LQR control problem:

$$\min_u J_{LQR} = \min_u \int_0^\infty \left[x_{LQR}^T(t) Q_{LQR} x_{LQR}(t) + R_{LQR} \delta^2(t) \right] dt$$

s.t.

$$\dot{x}_{LQR} = A_{path} x_{LQR} + B_{1,path} \delta + B_{2,path} k_{ref} \tag{19}$$

where $Q_{LQR} \in \mathbb{R}^{4\times4}$ and $R_{LQR} \in \mathbb{R}$ are positive weights (19) results into:

$$K_{LQR} = -R_{LQR}^{-1} B_{1,path}^T P \tag{20}$$

where P is the solution of the algebraic Riccati equation:

$$Q_{LQR} - P B_{1,path} R_{LQR}^{-1} B_{1,path}^T P^T + P A_{path} + A_{path}^T P = 0 \tag{21}$$

Similarly to the pole placement method, the feedforward gain $K_{ff,LQR}$ is tuned to have zero lateral displacement error in steady-state conditions. Moreover, the LQR problem is solved for different vehicle longitudinal speeds, and the resulting control gains are collected in look-up tables.

4.3.4 Sliding Mode Control

The general idea behind the sliding mode control methods is to use discontinuous control actions to create an attractive surface in the system state-space, known as the sliding surface. On the sliding surface, the system is controlled via an equivalent control action, which steers the system dynamics towards the desired trajectory or equilibrium point. Although the discontinuous control component makes the sliding surface attractive despite parameter uncertainties and unmodelled dynamics, it can also induce high-frequency variations in the control action, known as chattering, which could ultimately compromise the actuators. The SMC-PT solution, implemented in the control framework in Fig. 5, is based on the work presented in [29] and the following STM formulation:

$$\begin{aligned}\begin{bmatrix}\dot{\beta}\\ \ddot{\psi}\end{bmatrix} &= \begin{bmatrix}\frac{C_f+C_r}{mv_x} & \frac{L_fC_f-L_rC_r}{mv_x}-1\\ \frac{L_fC_f-L_rC_r}{I_z} & \frac{L_f^2C_f+L_r^2C_r}{I_zv_x}\end{bmatrix}\begin{bmatrix}\beta\\ \dot{\psi}\end{bmatrix} - \begin{bmatrix}\frac{C_f}{mv_x}\\ \frac{aC_f}{I_z}\end{bmatrix}\delta\\ &= \begin{bmatrix}A_I & A_{II}\\ A_{III} & A_{IV}\end{bmatrix}\begin{bmatrix}\beta\\ \dot{\psi}\end{bmatrix} - \begin{bmatrix}B_I\\ B_{II}\end{bmatrix}\delta\end{aligned} \tag{22}$$

The control input that ensures asymptotic stability and achieves the convergence of the state variables (namely the yaw rate error and the yaw rate acceleration error) towards zero is:

$$\delta = \frac{I_z}{C_f}\left(A_{III}\beta + A_{IV}\dot{\psi} + k_{smc,1}\left(\dot{\psi} - \dot{\psi}_{ref}\right)\right) - \frac{I_z}{L_fC_f}k_{smc,2}\text{sign}(\sigma_c) \tag{23}$$

where $k_{smc,1}$ and $k_{smc,2}$ are control parameters; $\dot{\psi}_{ref}$ is the reference yaw rate obtained from the reference trajectory; and σ_c is the sliding variable, defined as the linear combination of the error and its time integral. To limit the chattering of the control input, the sign function has been implemented as a sigmoid function. Furthermore, the control gains are varied with the longitudinal vehicle speed.

4.3.5 Proportional Integral and Stanley Controllers

The PI control law is defined as:

$$\delta = Lk_{ref} + K_p\Delta y + K_I\int_0^t \Delta y(\tau)d\tau \tag{24}$$

where Lk_{ref} is a feedforward term derived from the Ackermann steering equation, while K_p and K_I are the proportional and integral gains related to the feedback term Δy.

The proposed Stanley controller is based on the strategy presented in [30], which is commonly used as baseline for comparative analyses. STC is a nonlinear feedback function, whose control input is computed as:

$$\delta = \Delta\psi + \arctan\left(\frac{k_{stc}\Delta y}{v_x}\right) \tag{25}$$

where k_{stc} is a tuneable parameter.

5 Experimental Path Tracking Results

5.1 Considered Scenarios and Key Performance Indicators

The proposed PT strategies are evaluated along five paths, each encompassing eight constant levels of the reference longitudinal vehicle speed, i.e., $v_x \in \{0.5, 0.6, 0.7, 0.8, 0.9, 1, 1.2, 1.5\}$ m/s. The selected reference paths are illustrated in Fig. 6, and in the remainder are referred to as: (a) U-shape; (b) S-shape; (c) O-shape; (d) ∞-shape; and (e) obstacle avoidance shape (OAV-shape). Specifically, the OAV-shape is a manoeuvre consisting of a U-turn, followed by a collision avoidance manoeuvre for passenger cars, according to the ISO 3888 2 standard, with a factor 10 scaling to match the size of the scaled robotic vehicle.

The following set of KPIs is used to quantitatively evaluate the closed-loop behaviour provided by the six PT solutions:

- The maximum and root mean square values of the lateral error, ME and RMSE:

$$\mathrm{ME} = \max_{t\in[0,T_f]} |\Delta y(t)|$$

$$\mathrm{RMSE} = \sqrt{\frac{1}{T_f}\int_0^{T_f} |\Delta y(t)|^2 dt} \qquad (26)$$

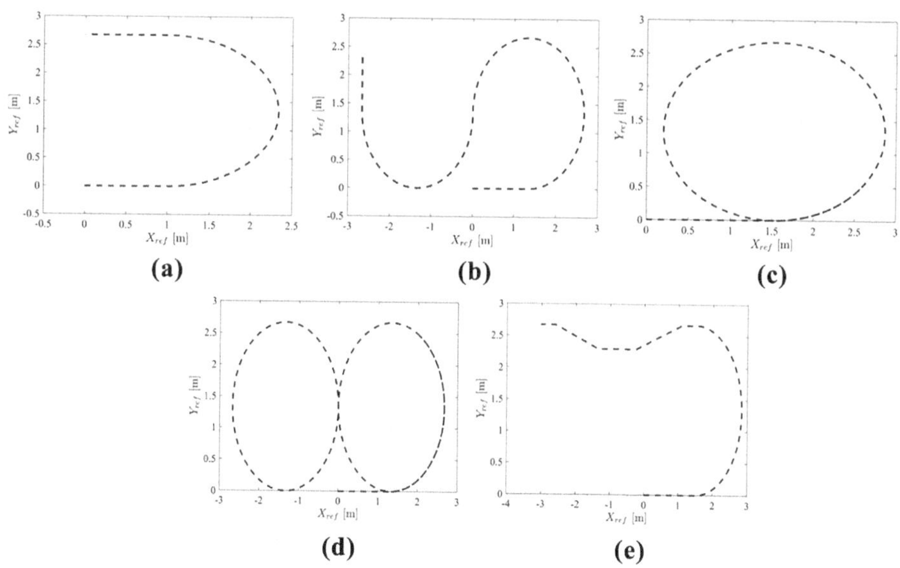

Fig. 6 Considered reference paths: **a** U-shape; **b** S-shape; **c** O-shape; **d** ∞-shape; and **e** OAV-shape

where T_f is the duration of the relevant part of the manoeuvre;

- The normalised integral of the absolute value of the steering angle control action (IACA):

$$\text{IACA} = \frac{1}{T_f} \int_0^{T_f} |\delta(t)| dt. \tag{27}$$

ME and RMSE are widely used to compare the closed-loop tracking performance of PT strategies [11], while IACA is a measurement of the control effort required by each PT controller.

5.2 Experimental Results

Figures 7, 8, 9, 10 and 11 show the experimental lateral error profiles along the travelled distance s, obtained for the (i) U-shape (Fig. 7); (ii) S-shape (Fig. 8); (iii) O-shape (Fig. 9); (iv) ∞-shape (Fig. 10); and (v) OAV-shape (Fig. 11) tests, for three speeds, i.e., $v_x = 0.5$, 1.0, and 1.5 m/s, while Figs. 12, 13, 14, 15 and 16 present the KPIs computed for the entire set of considered vehicle speeds.

From the analysis of the vehicle response profiles and KPI histograms, the following remarks can be drawn:

- At low speed, e.g., from 0.5 to 0.7 m/s, most of the proposed PT controllers exhibit similar closed-loop behaviour across all the manoeuvres (see Figs. 7, 8, 9, 10 and 11a). A reduced tracking performance is observed for STC with respect to the other strategies. For example, the ME and RMSE values are approximately twice those obtained for the PI case, which is characterised by similar control effort (IACA). A notable decline of the closed-loop tracking performance occurs during the OAV-shape manoeuvre, which has sharp variations of the road curvature, and thus requires substantial steering angle adjustments. In this scenario, the lateral error KPIs for STC increase by approximately four times compared to those of the PI implementation.

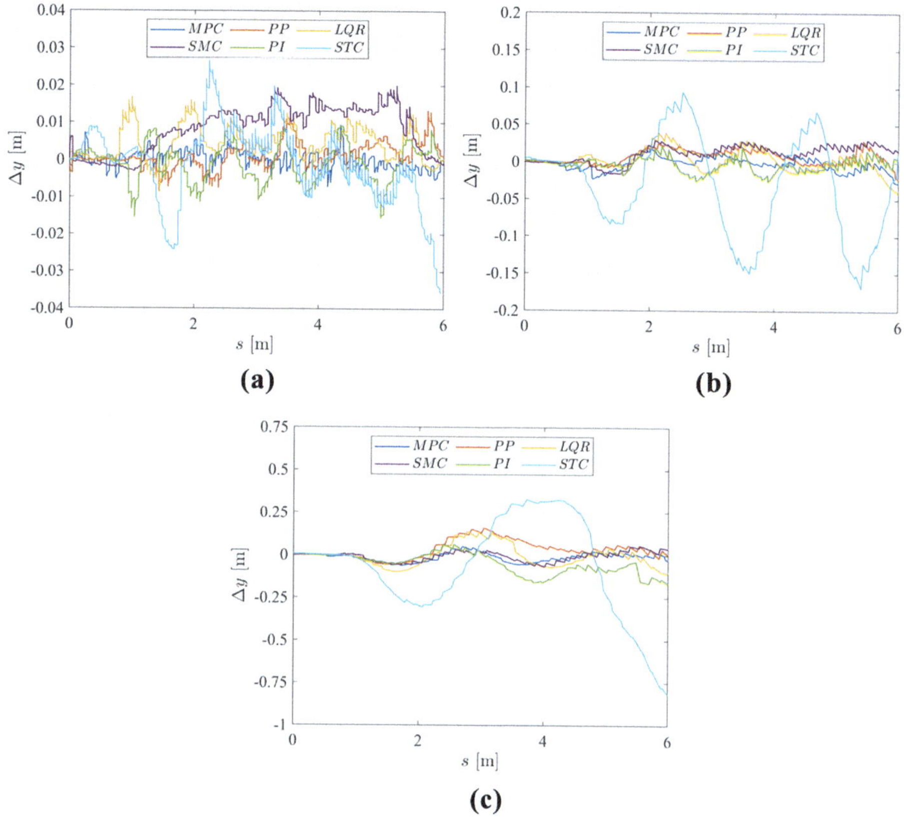

Fig. 7 Experimental lateral error dynamics provided by the considered control strategies in the U-shape manoeuvre at: **a** $v_x = 0.5$ m/s; **b** $v_x = 1$ m/s; and **c** $v_x = 1.5$ m/s

- When the speed is increased, the performance differences become more pronounced, with the STC showing markedly worse performance than the other PT approaches, and its lateral error reaching peaks ranging from 0.5 to 1 m. This is likely due to the simplistic nature of the Stanley controller, which operates as a geometric controller, and thus cannot be effective in conditions with significant cornering dynamics.
- At higher vehicle speeds, especially in the most challenging scenarios, i.e., the S-shape, ∞-shape, and OAV-shape tests, the PI controller struggles providing good PT control, as observed by the increased steering effort and the resulting poor trajectory tracking performance. However, for the least demanding manoeuvres (i.e., the O- and U-shape

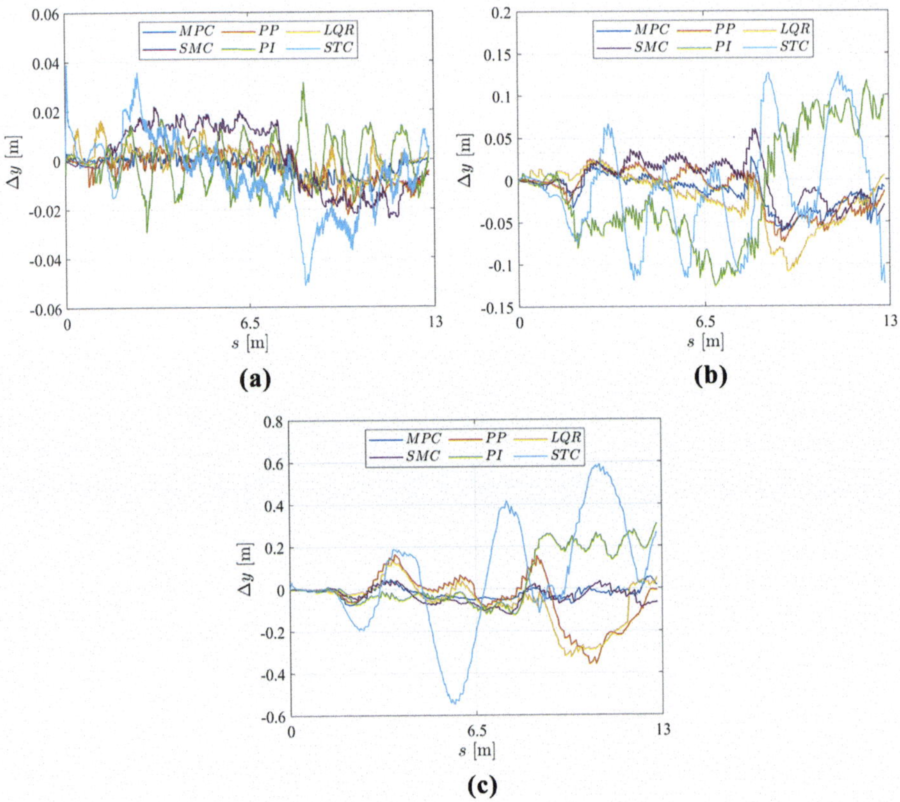

Fig. 8 Experimental lateral error dynamics provided by the considered control strategies in the S-shape manoeuvre at: **a** $v_x = 0.5$ m/s; **b** $v_x = 1$ m/s; and **c** $v_x = 1.5$ m/s

tests), the PI controller still provides similar closed-loop response to the PP and LQR strategies.

- Overall, the PP and LQR algorithms are characterised by: (i) a similar closed-loop behaviour, with lower lateral error peaks and a smoother response when compared to the PI and STC implementations, as observed in Figs. 7, 8, 9, 10 and 11; and (ii) equivalent performance in terms of lateral error, as confirmed by the corresponding KPIs.

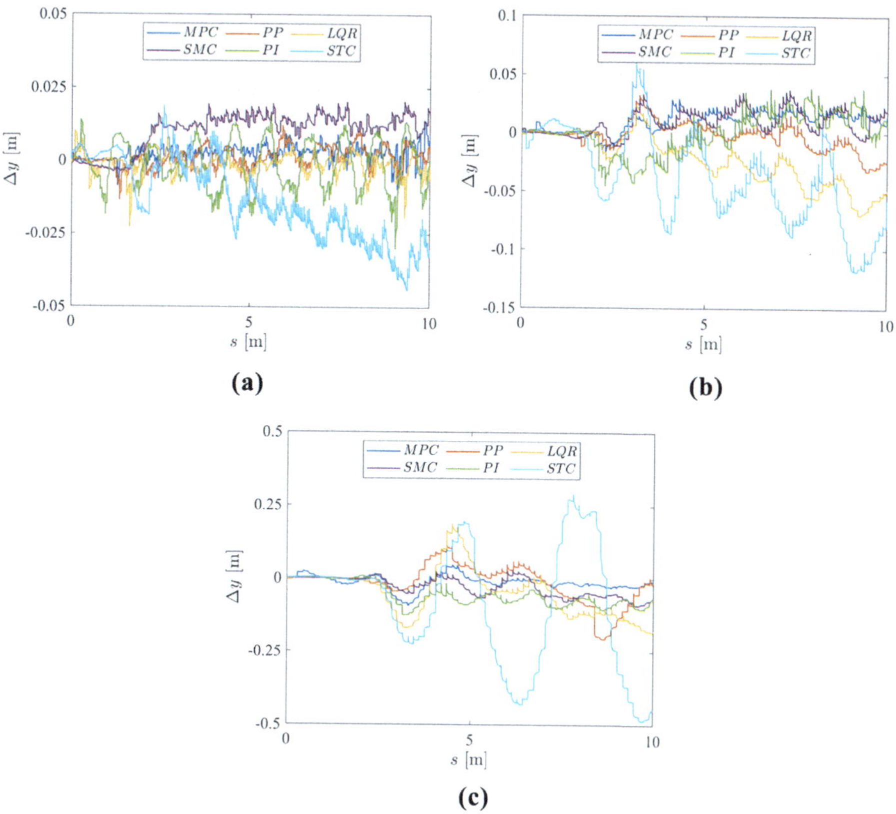

Fig. 9 Experimental lateral error dynamics provided by the considered control strategies in the O-shape manoeuvre at: **a** $v_x = 0.5$ m/s; **b** $v_x = 1$ m/s; and **c** $v_x = 1.5$ m/s

- MPC and SMC provide the best results among the proposed PT controllers, outperforming also LQR and PP, e.g., in terms of residual lateral error. More specifically, (i) MPC and SMC achieve the lowest ME and RMSE values, especially at high speed and in the most challenging scenarios, i.e., the S-shape, ∞-shape, and OAV-shape tests, where ME and RMSE are always kept below 0.2 and 0.1 m, respectively; and (ii) SMC requires relatively low control effort, with reduced IACA by at least ~20% compared to MPC.

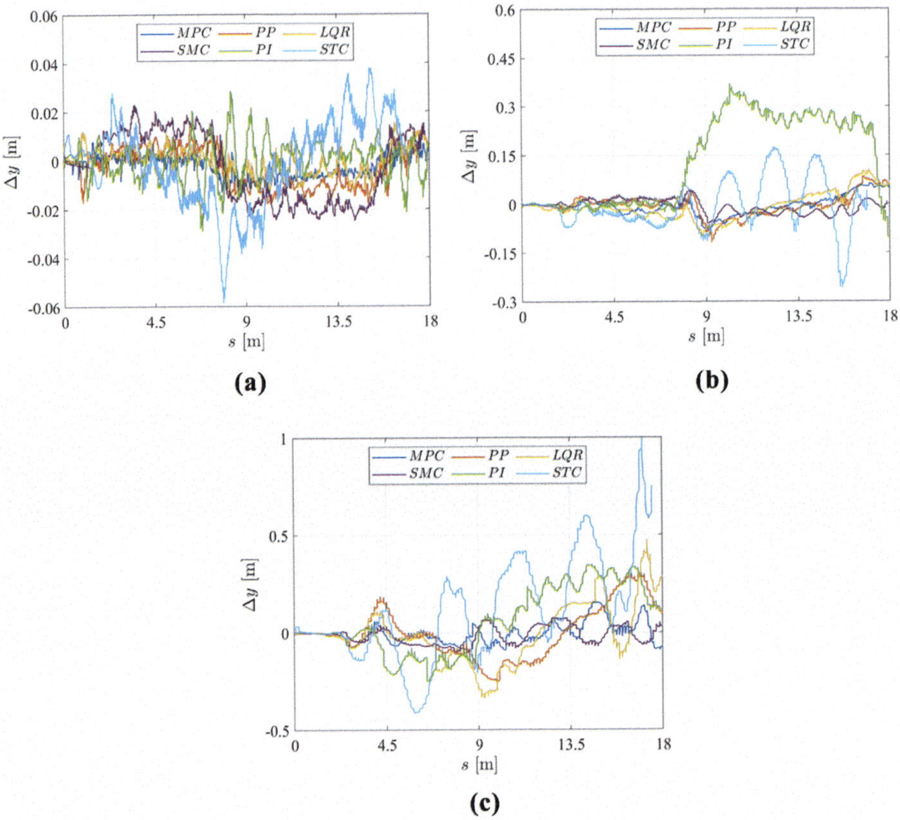

Fig. 10 Experimental lateral error dynamics provided by the considered control strategies in the ∞-shape manoeuvre at: **a** $v_x = 0.5$ m/s; **b** $v_x = 1$ m/s; and **c** $v_x = 1.5$ m/s

In summary, the analysis shows that the model-based approaches—MPC, PP, LQR, and SMC—outperform both the PI and STC strategies in all the proposed scenarios, due to their knowledge of the vehicle dynamics. In particular, two PT strategies stand out: (i) MPC, due to its ability to optimise system dynamics while considering constraints; and (ii) SMC, whose discontinuous control term enhances the robustness against the unknown dynamics, which become more significant during aggressive manoeuvres. These aspects provide a key advantage over the other control strategies, resulting in better KPIs especially at the highest vehicle speeds and for the most challenging paths.

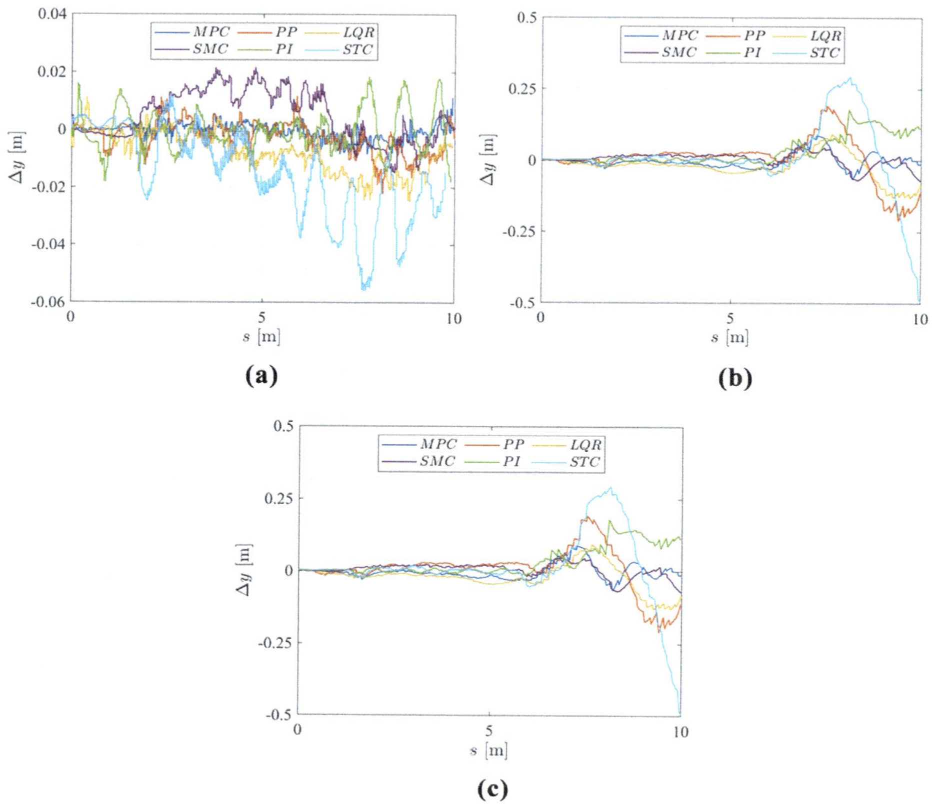

Fig. 11 Experimental lateral error dynamics provided by the considered control strategies in the *OAV*-shape manoeuvre at: **a** $v_x = 0.5$ m/s; **b** $v_x = 1$ m/s; and **c** $v_x = 1.5$ m/s

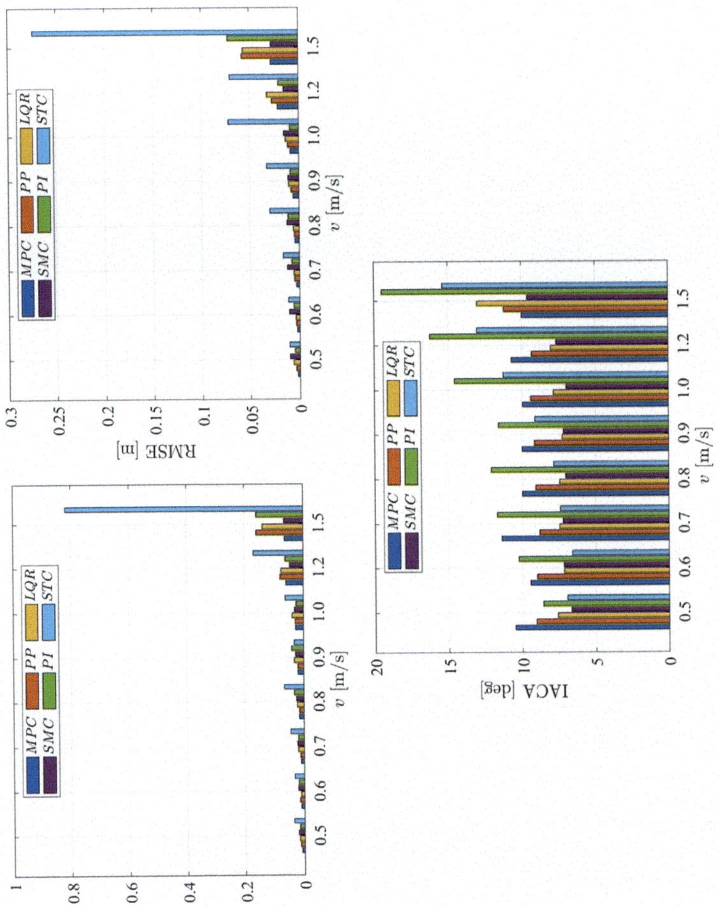

Fig. 12 Experimental KPIs for the *U*-shape manoeuvre

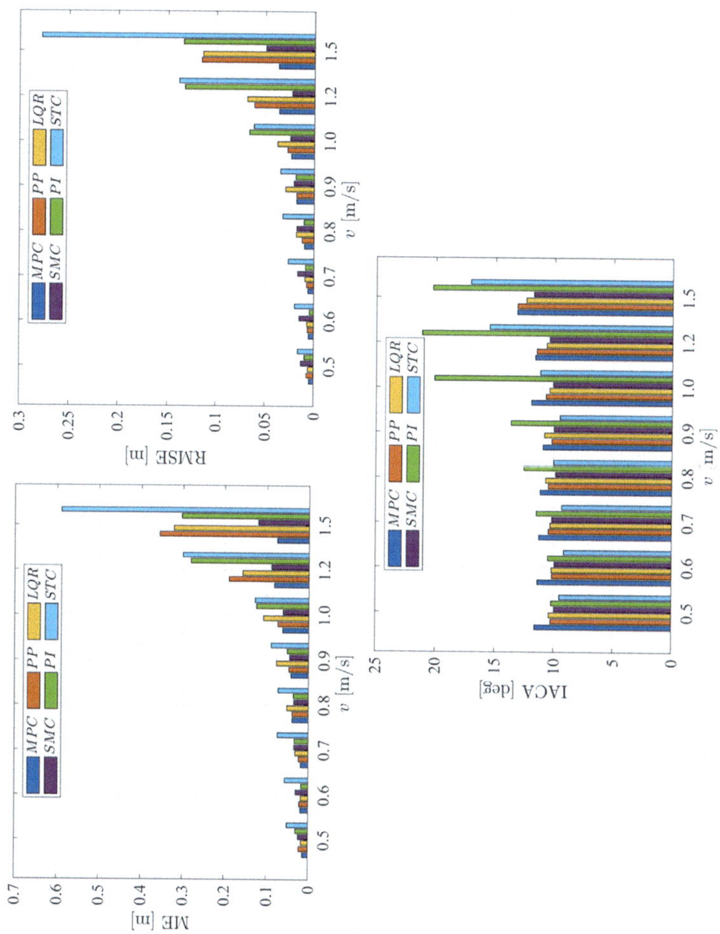

Fig. 13 Experimental KPIs for the *S*-shape manoeuvre

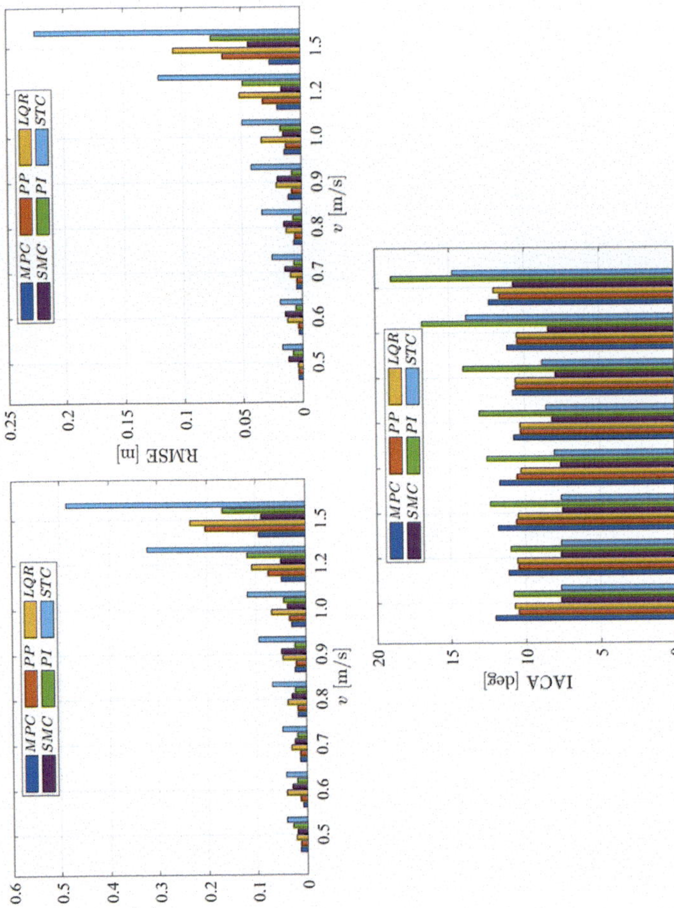

Fig. 14 Experimental KPIs for the O-shape manoeuvre

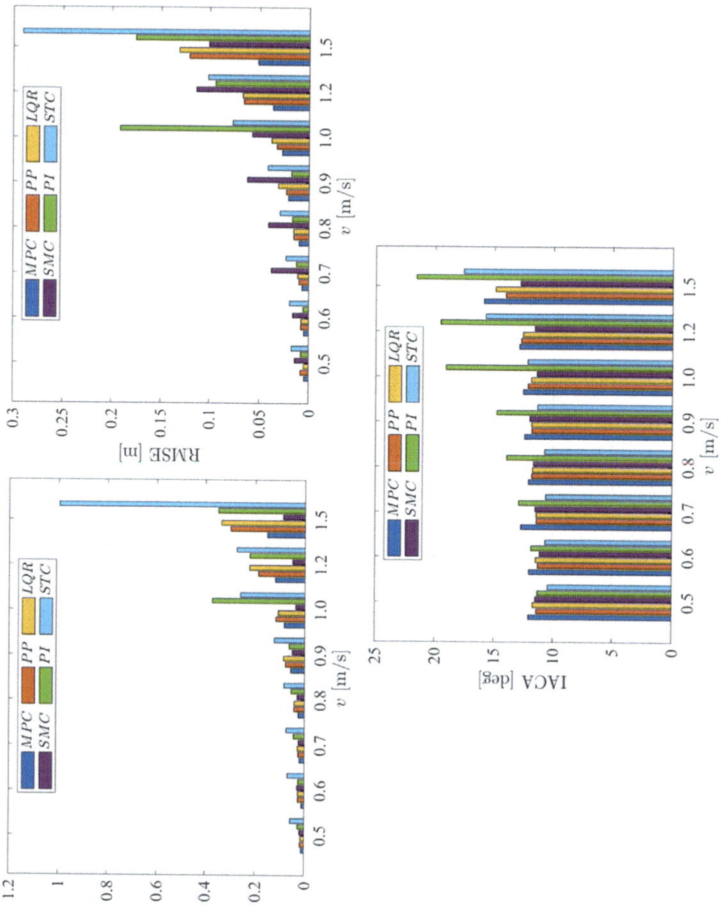

Fig. 15 Experimental KPIs for the ∞-shape manoeuvre

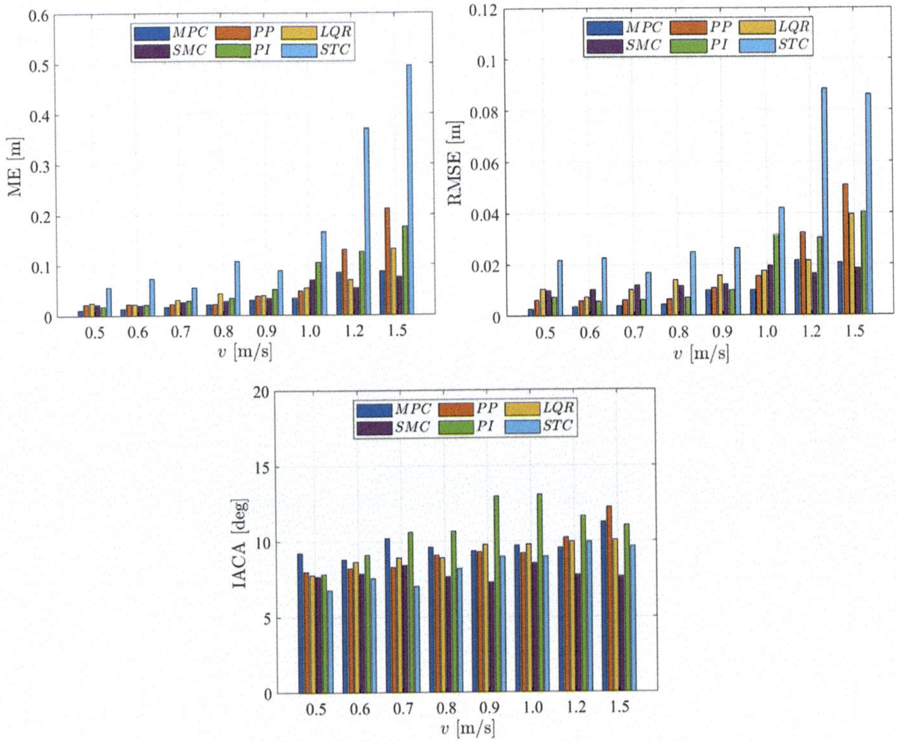

Fig. 16 Experimental KPIs for the *OAV*-shape manoeuvre

6 Conclusion

This study presented the formulation, implementation, and experimental comparison of six path tracking (PT) controllers for automated scaled vehicles, by using a set of key performance indicators. The algorithms include a model predictive control (MPC) implementation, a pole placement (PP) formulation, a linear quadratic regulator (LQR), a sliding mode control (SMC) strategy, a proportional integral (PI) controller, and Stanley control (STC) strategy. An extensive experimental campaign was carried out along five manoeuvres at eight vehicle speeds. The ability of MPC of achieving the online solution of a constrained optimal control problem embedding the vehicle dynamics as well as the robustness provided by the discontinuous SMC control action are key advantages, bringing superior closed-loop tracking performance. In fact, the results show that MPC and SMC outperform the other strategies in terms of reductions of the maximum and root mean square values of the lateral error, with an acceptable level of control effort. Such performance benefit becomes more evident when the vehicle is operated under challenging conditions, i.e., at the highest speeds of the selected range and for the reference paths with sharp road curvature variations.

References

1. Huang, J., Junginger, S., Liu, H., Thurow, K.: Indoor positioning systems of mobile robots: a review. Robotics **12**(2) (2023)
2. Lynch, L., Newe, T., Clifford, J., Coleman, J., Walsh, J., Toal, D.: Automated ground vehicle (AGV) and sensor technologies—a review. In: 12th International Conference on Sensing Technology (ICST), pp. 347–352 (2018)
3. Rupp, A., Tranninger, M., Wallner, R., Zubaca, J., Steinberger, M., Horn, M.: Fast and low-cost testing of advanced driver assistance systems using small-scale vehicles. In: IFAC-PapersOnLine, 9th IFAC Symposium on Advances in Automotive Control AAC 2019, vol. 52, no. 5, pp. 34–39 (2019)
4. Ribeiro, A., Koyama, M., Moutinho, A., de Paiva, E., Fioravanti, A.: A comprehensive experimental validation of a scaled car-like vehicle: lateral dynamics identification, stability analysis, and control application. Control Eng. Pract. **116**, 104924 (2021)
5. Liu, L., Wang, X., Yang, X., Liu, H., Li, J., Wang, P.: Path planning techniques for mobile robots: review and prospect. Expert Syst. Appl. **227**, 120254 (2023)
6. Hu, J., Zhang, Y., Rakheja, S.: Adaptive lane change trajectory planning scheme for autonomous vehicles under various road frictions and vehicle speeds. IEEE Trans. Intell. Veh. **8**(2), 1252–1265 (2023)
7. Tzafestas, S.G.: Mobile robot control and navigation: a global overview. J. Intell. Rob. Syst. **91**, 35–58 (2018)
8. Li, Q., Chen, Z., Li, X.: A review of connected and automated vehicle platoon merging and splitting operations. IEEE Trans. Intell. Transp. Syst. **23**(12), 22790–22806 (2022)
9. Dixit, S., Fallah, S., Montanaro, U., Dianati, M., Stevens, A., Mccullough, F., Mouzakitis, A.: Trajectory planning and tracking for autonomous overtaking: state-of-the-art and future prospects. Annu. Rev. Control **45**, 76–86 (2018)
10. Montanaro, U., Dixit, S., Fallah, S., Dianati, M., Stevens, A., Oxtoby, D., Mouzakitis, A.: Towards connected autonomous driving: review of use-cases. Veh. Syst. Dyn. **57**(6), 779–814 (2018)
11. Stano, P., Montanaro, U., Tavernini, D., Tufo, M., Fiengo, G., Novella, L., Sorniotti, A.: Model predictive path tracking control for automated road vehicles: a review. Annu. Rev. Control **55**, 194–236 (2023)
12. Dixit, S., Montanaro, U., Dianati, M., Mouzakitis, A., Fallah, S.: Integral MRAC with bounded switching gain for vehicle lateral tracking. IEEE Trans. Control Syst. Technol. **29**(5), 1936–1951 (2021)
13. Rokonuzzaman, M., Mohajer, N., Nahavandi, S., Mohamed, S.: Review and performance evaluation of path tracking controllers of autonomous vehicles. IET Intel. Transp. Syst. **15**(5), 646–670 (2021)
14. Hung, N., Rego, F., Quintas, J., Cruz, J., Jacinto, M., Souto, D., Potes, A., Sebastiao, L., Pascoal, A.: A review of path following control strategies for autonomous robotic vehicles: theory, simulations, and experiments. J. Field Robot. **40**(3), 747–779 (2023)

15. Bryan, W.T., Boler, M.E., Bevly, D.M.: A vehicle-independent autonomous lane keeping and path tracking system. IFAC PapersOnLine **54**(2), 37–44 (2021)
16. Chu, Z., Sun, Y., Wu, C., Sepehri, N.: Active disturbance rejection control applied to automated steering for lane keeping in autonomous vehicles. Control Eng. Pract. **74**, 13–21 (2018)
17. Karimi Pour, F., Theilliol, D., Puig, V., Cembrano, G.: Health aware control design based on remaining useful life estimation for autonomous racing vehicle. ISA Trans. **113**, 196–209 (2021)
18. Liniger, A., Lygeros, J.: Real-time control for autonomous racing based on viability theory. IEEE Trans. Control Syst. Technol. **27**(2), 464–478 (2019)
19. Alcala, E., Puig, V., Quevedo, J., Rosolia, U.: Autonomous racing using linear parameter varying-model predictive control (LPV-MPC). Control Eng. Pract. **95**, 104270 (2020)
20. Artunedo, A., Moreno-Gonzalez, M., Villarga, J.: Lateral control for autonomous vehicles: a comparative evaluation. Annu. Rev. Control. **57**, 100910 (2024)
21. Quanser: QCar (2023). https://www.quanser.com/products/qcar/. Accessed 14 September 2023
22. Caponio, C., Stano, P., Carli, R., Olivieri, I., Ragone, D., Sorniotti, A., Gruber, P., Montanaro, U.: Modeling, positioning, and deep reinforcement learning path following control of scaled robotic vehicles: design and experimental validation. IEEE Trans. Autom. Sci. Eng. (2025) (in press)
23. The MathWorks Inc.: Matlab version: 9.13.0 (r2022b) (2022). https://www.mathworks.com. Accessed 14 September 2023
24. Chatzikomis, C., Sorniotti, A., Gruber, P., Zanchetta, M., Willans, D., Balcombe, B.: Comparison of path tracking and torque-vectoring controllers for autonomous electric vehicles. IEEE Trans. Intell. Veh. **3**(4), 559–570 (2018)
25. Rajamani, R.: Vehicle Dynamics and Control. Springer (2006)
26. Grune, L., Pannek, J.: Nonlinear Model Predictive Control. Springer (2011)
27. Houska, B., Ferreau, H., Diehl, M.: ACADO toolkit—an open source framework for automatic control and dynamic optimization. Optim. Control Appl. Methods **32**(3), 298–312 (2011)
28. Schmeitz, A., Zegers, J., Ploeg, J., Alirezaei, M.: Towards a generic lateral control concept for cooperative automated driving theoretical and experimental evaluation. In: 5th IEEE International Conference on Models and Technologies for Intelligent Transportation Systems (MT-ITS), pp. 134–139 (2017)
29. Hu, H., Bei, S., Zhao, Q., Han, X., Zhou, D., Zhou, X., Li, B.: Research on trajectory tracking of sliding mode control based on adaptive preview time. Actuators **11**(2), 34 (2022)
30. Hoffmann, G.M., Tomlin, C.J., Montemerlo, M., Thrun, S.: Autonomous automobile trajectory tracking for off-road driving: Controller design, experimental validation and racing. In: American Control Conference, pp. 2296–2301 (2007)

A Nonlinear Dead-Time Compensation Method for Path Tracking Control

Karin Festl and Michael Stolz

Abstract

In the realm of autonomous vehicle technologies and advanced driver assistance systems, precise and reliable path tracking controllers are vital for safe and efficient navigation. However the presence of dead time in the vehicle control systems poses a challenge to real-world systems. Input and output delays are caused by factors like sensor processing and mechanical response and can range up to a few hundred milliseconds. This chapter addresses the problem of dead time in path tracking control and proposes a method to compensate the dead time. The proposed solution involves a nonlinear prediction model, in a structure similar to the Smith predictor, but incorporating the kinematic behavior of the vehicle plant system. The implementation avoids numeric integration or optimization, enabling a fast execution. Simulation tests with various controllers and disturbances, including dead-time uncertainty, demonstrate the efficacy of the dead-time compensation method. Results indicate improved control performance in all tested scenarios.

Keywords

Dead-time compensator • Path tracking • Nonlinear control

K. Festl (✉) · M. Stolz
Virtual Vehicle Research GmbH, Graz, Austria
e-mail: karin.festl@v2c2.at

D. Watzenig (ed.), *Recent Advances in Autonomous Vehicle Technology–Perception and Path Planning*, Sustainable Mobility & Energy,
https://doi.org/10.1007/978-3-032-01001-8_7

7.1 Introduction

In the area of autonomous vehicle technologies and advanced driver assistance systems, the precision and reliability of path tracking controllers play a pivotal role in ensuring safe and efficient navigation. However, real-world systems often face challenges, one of which is the presence of dead time. The dead time composes of both a delay between the output of a steering command and its actual execution (output delay) and of the time that passes until a state feedback value is generated from the actual vehicle state (input delay). Dead time can result from various factors, including sensor processing, communication latency, and mechanical response.

Addressing dead time is crucial for achieving accurate and robust vehicle control. For many advanced control strategies, as for example optimal predictive control, dead-time compensation is intrinsic. However, in dynamic environments where rapid decision-making and execution is essential, faster control algorithms are used, which often require additional modifications to compensate dead time. The Smith predictor is a well known method for compensating dead time. In its simplest form, it is not applicable to unstable plants or plants with an integrator. With modifications such as the Smith-Åström predictor [5], the dead-time compensation can be applied on a broad range of control systems. Similar approaches are also applied to the path tracking control problem: In [2] an observer similar to Luenberger observer is applied to the nonlinear plant system. In [7], an optimal predictor and a Smith predictor are applied to the linearized vehicle model plant, comparing their robustness.

In this chapter, we propose a nonlinear method for dead-time compensation that modifies the state feedback similar to the Smith predictor, however, by incorporating the kinematic behavior of the path tracking plant system, it is possible to apply the predictive correction directly to the unstable nonlinear plant system. While this method is based on model state prediction, it does not involve numeric integration or optimization but can be implemented in explicit form. We test the dead-time compensation in simulation with different controllers and in presence of different kinds of disturbances and uncertainties, including dead-time uncertainty. The results show that the dead-time compensation improves the control performance in all tested scenarios.

The chapter starts by defining the dead-time problem and the desired behavior of the system with dead-time compensation. After that, the solution is derived and demonstrated in a short example.

7.2 Problem Definition

Suppose there is a path tracking controller that shows the desired performance for a given use case, when there is no dead time in the system. When dead time emerges in the system, the controller adaptions are required. We first describe the control system comprising dead time and afterwards define the desired properties of a dead-time compensation method.

7.2.1 Path Tracking with Dead Time

The path tracking (or path stabilization) problem can be described as controlling the steering of a vehicle such that it approaches and follows a reference or target path. More formally, we need to find a feedback law $\delta\left(\boldsymbol{p}, \boldsymbol{p}_{\text{ref}}\right)$ such that

$$\exists s : \lim_{t \to \infty} ||\boldsymbol{p}(t) - \boldsymbol{p}_{\text{ref}}(s)|| \leq \varepsilon \tag{7.1}$$

where $\boldsymbol{p}(t)$ is the vehicle reference point at time t, $\boldsymbol{p}_{\text{ref}}(s)$ is the reference path (we define the path as parametric curve with s being the length along the curve) and ε is a tolerance distance. In other words, we require a feedback law that steers the minimum distance between vehicle and reference path to a vicinity around 0. Figure 7.1 illustrates the vehicle with reference point $\boldsymbol{p}$ and the reference path $\boldsymbol{p}_{\text{ref}}(s)$.

In the simplest case, the vehicle dynamics $d/dt\ \boldsymbol{p}(t) = \dot{\boldsymbol{p}}(t)$ can be modelled by a kinematic single track model:

$$\boldsymbol{x} = \left[p_x\ p_y\ \psi\right]^T = \left[\boldsymbol{p}\ \psi\right]^T \tag{7.2a}$$

$$\dot{\boldsymbol{x}} = \begin{bmatrix} \cos\psi \\ \sin\psi \\ \tan(\delta)/l \end{bmatrix} \cdot v \tag{7.2b}$$

where δ is the steering angle, v is the velocity and l is the wheelbase of the vehicle. The real vehicle behavior will differ from this model due to several effects which may or may not be considered by the controller. Regardless, the dead-time compensation method will be based on this model. Moreover, we will formulate the approaches in discrete time: we denote with ξ_i any variable ξ at time $t = i \cdot \Delta t$, where Δt is the constant sample time. The dynamic equations of the kinematic single track model in presence of input delay Δt_{I} and output delay Δt_{O} is:

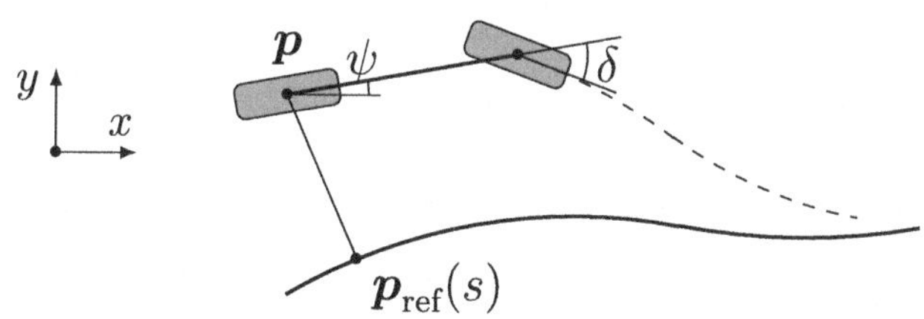

Fig. 7.1 Vehicle with reference point $\boldsymbol{p}$ in the rear axle approaching a reference path $\boldsymbol{p}_{\text{ref}}(s)$

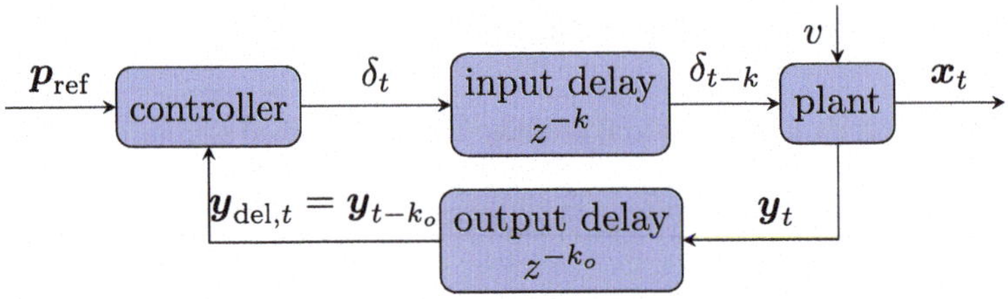

Fig. 7.2 Structure of the controlled plant with input and output delay

$$x_{del,i+1} = x_{del,i} + \int_{i\cdot\Delta t}^{(i+1)\cdot\Delta t} \begin{bmatrix} \cos\varphi(\tau) \\ \sin\varphi(\tau) \\ \tan(\delta_{i-k_I})/l \end{bmatrix} \cdot v_i \cdot d\tau \tag{7.3a}$$

$$y_{del,i} = y_{i-k_O} = x_{del,i} \cdot q^{-k_O} \tag{7.3b}$$

where y_{del} is the delayed state feedback, $k_I = \Delta t_I/\Delta t$, $k_O = \Delta t_O/\Delta t$ are the discrete delays and q is the forward shift operator as defined in [3]: $q \cdot \xi_i = \xi_{i+1}$. In Fig. 7.2, the structure of the controlled planned in the presence of input and output delay is shown.

We will demonstrate our approaches on three different path tracking controllers:

- Stanley controller [4]:
$$\delta = \varphi - \varphi_{ref} + \arctan\left(\frac{k \cdot e}{v}\right)$$
where e is the shortest distance between vehicle front axle and the reference path: $e = \min_s ||p - p_{ref}(s)||$, φ_{ref} is the orientation of the reference path at the point s of shortest distance and k is a tuning parameter.
- Pure pursuit controller [1]:
$$\delta = \arctan\left(\frac{2 \cdot l \cdot e_{pp}}{l_h^2}\right)$$
where e_{pp} is the tracking error in a lookahead distance l_h in front of the vehicles rear axle and l is the vehicles wheelbase.
- Dubins robust control [9]:
$$\delta = \bar{\delta} \cdot \text{sign}(\sigma)$$
where $\sigma(e, (\varphi - \varphi_{ref}), k_{rob})$ is a switching function implementing Dubins shortest path for reaching a reference path and k_{rob} is a tuning parameter.

These controllers differ in their sensitivity to dead time. To demonstrate this, we simulate the controllers with the plant (7.3). The vehicle is driving at a speed of 1 m/s so that the kinetic effects are small and the control performance is only influenced by the dead time. We test the controllers with dead time ($\Delta t_{in} + \Delta t_{out} = 0.4$ s) and without ($\Delta t_{in} = \Delta t_{out} = 0$ s).

The controllers are simulated at a frequency of 100 Hz ($\Delta t = 0.01$ s). The controllers are parameterized as:

$$k = 3.0\,\mathrm{s}^{-1}$$
$$l_\mathrm{h} = 1.45\,\mathrm{m},\ l = 1.0\,\mathrm{m}$$
$$k_\mathrm{rob} = 0.5$$

Dubins robust control parameter k_rob has been chosen in the centre of the permissible range. The Stanley control parameter k has been chosen such that the reaching behavior without dead-time resembles the one of Dubins robust control, to increase comparability. The lookahead distance l_h has been chosen such that the region of convergence is large enough for the starting position.

In Fig. 7.3, the paths produced by the different controllers are shown. In all 3 cases, the dead time leads to an oscillation around the reference path $\boldsymbol{p}_\mathrm{ref} = \left[x(s)\ 0\right]^T$. The lookahead of the pure pursuit controller, acts as a dead-time reduction. However, a large lookahead leads to a slow reaching behavior. A similar behavior is also true for the other controllers: when they are tuned to reach the reference path faster, then the dead time has a greater impact on the tracking performance. To demonstrate this, we simulate the Stanley controller with different values of the parmeter k. The results are shown in Fig. 7.4.

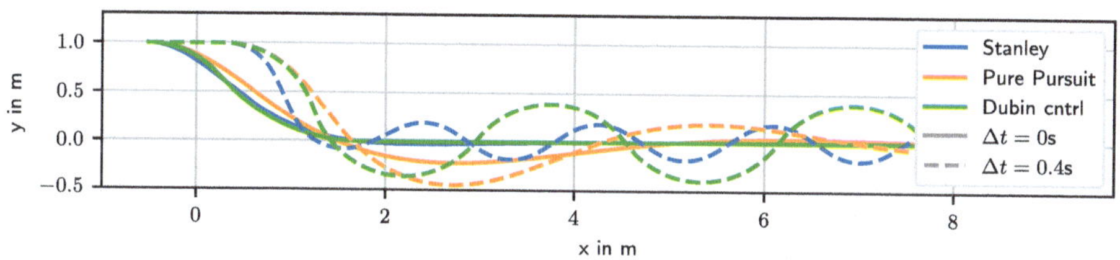

Fig. 7.3 Paths of three different controllers approaching and tracking the reference path $\boldsymbol{p}_\mathrm{ref} = [x(s)\,0]^T$. The plant is simulated first without dead time and then with a dead time of $\Delta t = 0.4$s

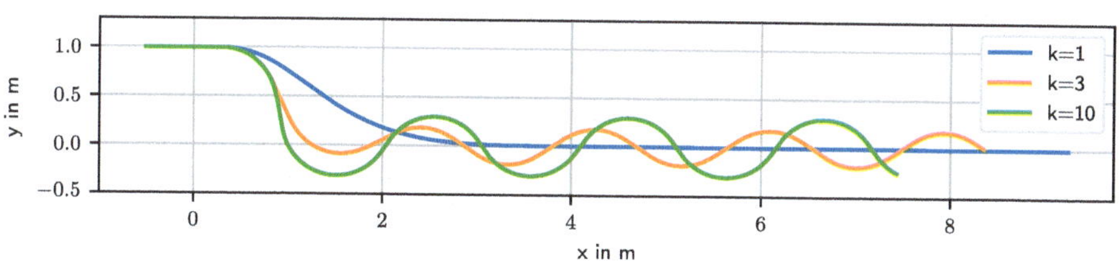

Fig. 7.4 Paths of the Stanley controller approaching and tracking the reference path $\boldsymbol{p}_\mathrm{ref} = [x(s)\,0]^T$. The plant has a dead time of $\Delta t = 0.4$ s and the controller is parametrized with different values for k

7.2.2 Dead-Time Compensation

As demonstrated, the controller typically can be tuned to more robustness to dead time in the plant system by decreasing the feedback gain or increasing a lookahead distance. However, this will decrease the reaching behavior and noise suppression. This robustness trade off is also discussed in [9]. Different to this approach, we want to compensate the dead time. With the use of a state prediction, it is possible to reduce the effect of dead time in the controlled loop without changing the controllers behavior. In short, we want to modify the plant such that it acts as if it had no dead time. In the following, we will give a mathematical formulation of this goal.

Mathematically, the plant model (7.3) implements a mapping from the system state and the steering input to the output. We define:

$$\mathbf{y}_{\text{del}} = f_P(\mathbf{x}, \delta \cdot q^{-k_\text{I}}) \cdot q^{-k_\text{O}} \tag{7.4}$$

When v is constant, then f_P is time invariant. In this case, the effect of input delay equals the effect of output delay. We define $k = k_\text{I} + k_\text{O}$:

$$\mathbf{y}_{\text{del}} = f_P(x, \delta \cdot q^{-k_\text{I}}) \cdot q^{-k_\text{O}} = f_P(x, \delta) \cdot q^{-k} \tag{7.5}$$

The goal of the dead-time compensation f_{dc} is to modify the feedback $\mathbf{y}_{\text{del}}$ such that the plant acts as if it had no delay:

$$\hat{\mathbf{y}} = f_{\text{dc}}(f_P(\mathbf{x}, \delta) \cdot q^{-k}, \delta) = f_{\text{dc}}(\mathbf{y}_{\text{del}}, \delta) \stackrel{!}{=} f_P(\mathbf{x}, \delta) \tag{7.6}$$

The structure of the control system with dead-time compensation is shown in Fig. 7.5.

Due to the time invariance (7.5), the dead-time compensation can be condensed to the task of predicting $\mathbf{y}$:

$$\hat{\mathbf{y}} = f_{\text{dc}}(\mathbf{y}_{\text{del}}, \delta) \stackrel{!}{=} \mathbf{y}_{\text{del}} \cdot q^{k} \tag{7.7}$$

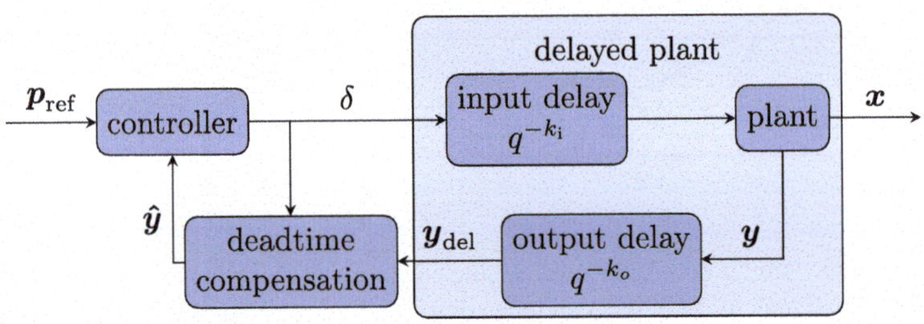

Fig. 7.5 Structure of the controlled system with dead time in the plant and a dead-time compensation to modify the control feedback $\mathbf{y}_{\text{del}}$

7.3 Nonlinear Motion Prediction

For predicting the undelayed plant output $\hat{\boldsymbol{y}}$ we use a plant model to compute future states. The plant model is a kinematic singletrack model (7.2). When the controller is implemented in discrete time, the steering angle δ is constant during one timestep. The plant model resembles the discrete formulation of the plant (7.3) without dead time, thus yielding the future state predictions $\hat{\boldsymbol{y}} \approx \boldsymbol{y}_{\text{del}} \cdot q^k$. With $\dot{\psi} = \tan(\delta)/l \cdot v$ we can write:

$$\hat{\boldsymbol{p}}_{i+1} - \hat{\boldsymbol{p}}_i = \int_0^{\Delta t} \begin{bmatrix} \cos(\hat{\psi}_i + \dot{\psi}_i \cdot \tau) \\ \sin(\hat{\psi}_i + \dot{\psi}_i \cdot \tau) \end{bmatrix} \cdot v \cdot d\tau = \int_0^{\Delta t} \boldsymbol{R}(\hat{\psi}_i) \cdot \begin{bmatrix} \cos(\dot{\psi}_i \cdot \tau) \\ \sin(\dot{\psi}_i \cdot \tau) \end{bmatrix} \cdot v \cdot d\tau \tag{7.8a}$$

$$= \boldsymbol{R}(\hat{\psi}_i) \cdot \begin{bmatrix} \sin(\dot{\psi}_i \cdot \Delta t) \\ (1 - \cos(\dot{\psi}_i \cdot \Delta t)) \end{bmatrix} \cdot \frac{v}{\dot{\psi}_i}$$

$$\hat{\psi}_{i+1} = \hat{\psi}_i + \dot{\psi} \cdot \Delta t \tag{7.8b}$$

$$\hat{\boldsymbol{y}}_i = \hat{\boldsymbol{x}}_i = \begin{bmatrix} \hat{\boldsymbol{p}}_i & \hat{\psi}_i \end{bmatrix}^T \tag{7.8c}$$

where $\boldsymbol{R}$ is the rotation matrix:

$$\boldsymbol{R}(\psi) = \begin{bmatrix} \cos(\psi) & -\sin(\psi) \\ \sin(\psi) & \cos(\psi) \end{bmatrix} \tag{7.9}$$

The plant model (7.8) is not internally stable (the state $\hat{\boldsymbol{x}}$ and output $\hat{\boldsymbol{y}}$ are ramp functions at steady state). In this case, a Smith predictor without modification is not applicable as $\hat{\boldsymbol{y}}$ will diverge [6]. There are solutions to regain internal stability as explained in [5]. In this chapter, we will derive a prediction model that enables a suitable approximation of the future state despite divergence of $\hat{\boldsymbol{y}}$. This model is then modified to gain internal stability.

Independent of divergence of $\hat{\boldsymbol{y}}$, for small k[1] we can state:

$$\hat{\boldsymbol{p}}_i - \hat{\boldsymbol{p}}_{i-k} = \sum_{j=i}^{i+k} \boldsymbol{R}(\hat{\psi}_{j-k}) \cdot \begin{bmatrix} \sin(\dot{\psi}_{j-k} \cdot \Delta t) \\ 1 - \cos(\dot{\psi}_{j-k} \cdot \Delta t) \end{bmatrix} \cdot \frac{v}{\dot{\psi}_{j-k}} \tag{7.10a}$$

$$\boldsymbol{p}_{i+k} - \boldsymbol{p}_i \doteq \sum_{j=i}^{i+k} \boldsymbol{R}(\psi_j) \cdot \begin{bmatrix} \sin(\dot{\psi}_{j-k} \cdot \Delta t) \\ 1 - \cos(\dot{\psi}_{j-k} \cdot \Delta t) \end{bmatrix} \cdot \frac{v}{\dot{\psi}_{j-k}} \tag{7.10b}$$

$$\hat{\psi}_i - \hat{\psi}_{i-k} = \sum_{j=i}^{i+k} \dot{\psi}_{j-k} \cdot \Delta t \doteq \psi_{i+k} - \psi_i \tag{7.10c}$$

---

[1] The error between the real vehicle dynamics and the stated kinematic approximation will accumulate over the steps k. Thus, for large k, the approximation will become inacurate.

From (7.10c) follows for $j \in [i, i+k]$:

$$\psi_i - \hat{\psi}_{i-k} \doteq \psi_j - \hat{\psi}_{j-k} \tag{7.11}$$

Thus we can conclude:

$$\begin{aligned} \boldsymbol{R}(\psi_i - \hat{\psi}_{i-k}) \cdot (\hat{\boldsymbol{p}}_i - \hat{\boldsymbol{p}}_{i-k}) &\doteq \sum_{j=i}^{i+k} \boldsymbol{R}(\psi_j - \hat{\psi}_{j-k} + \hat{\psi}_{j-k}) \cdot \begin{bmatrix} \sin(\dot{\psi}_{j-k} \cdot \Delta t) \\ 1 - \cos(\dot{\psi}_{j-k} \cdot \Delta t) \end{bmatrix} \cdot \frac{v}{\dot{\psi}_{j-k}} \\ &\doteq \boldsymbol{p}_{i+k} - \boldsymbol{p}_i, \end{aligned} \tag{7.12}$$

which leads to the state prediction:

$$\boldsymbol{x}_{i+k} \doteq \boldsymbol{x}_i + \begin{bmatrix} \boldsymbol{R}(\psi_i - \hat{\psi}_{i-k}) \\ 1 \end{bmatrix}^T \cdot (\hat{\boldsymbol{x}}_i - \hat{\boldsymbol{x}}_{i-k}) \tag{7.13}$$

This matches the structure of the Smith predictor except for the rotational transformation. The structure is also shown in Fig. 7.6. However, keep in mind that the underlying assumptions are different to the Smith predictor and consequently the behavior is different. The Smith predictor requires a stable plant such that the plant model state $\hat{\boldsymbol{x}}$ will not diverge. In our approach, we apply the dead-time compensation on an unstable plant, resulting in a possible divergence of $\hat{\boldsymbol{x}}$. Due to the special structure of the plant model, we can transform the prediction $\Delta\hat{\boldsymbol{x}} = \hat{\boldsymbol{x}}_i - \hat{\boldsymbol{x}}_{i-k}$ to account for the difference $\hat{\boldsymbol{x}}_{i-k} - \boldsymbol{x}_i$. This concept is illustrated in Fig. 7.7. The position prediction $\Delta\hat{\boldsymbol{p}}$ is transformed from the prediction model coordinate frame $\hat{T}$ to the vehicle coordinate frame T.

To make the behavior of the dead-time compensation more clear, we apply it on an ideal plant where the dead time is exactly known. The vehicle is steered to the reference path by the Dubins robust controller [9]. As there are no unkown effects, the reaching path with compensated dead time $\boldsymbol{p}(\Delta t_{\text{del}} = 0s)$ matches exactly the reaching path with dead time $\boldsymbol{p}(\Delta t_{\text{del}} = 1s)$, except that the approaching sequence starts at a delay equal to the dead time. This is shown in Fig. 7.8. Additionally shown in this figure is the position of the plant model $\hat{\boldsymbol{p}}$ of the dead-time compensator. The initial model state is different to the real position and it does not converge to the real position at any time. However, the shape of the plant

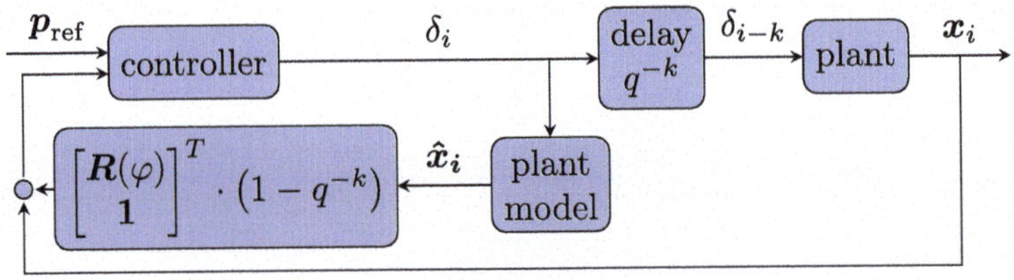

Fig. 7.6 Structure of the control system with a delayed plant and dead-time compensation with the help of a prediction plant model

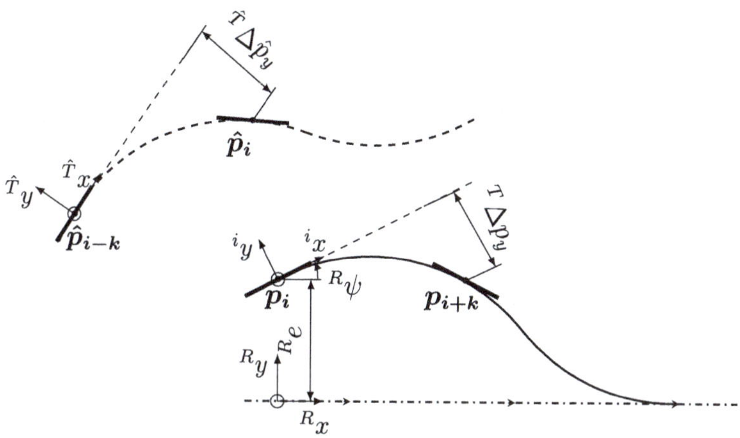

Fig. 7.7 Concept of the presented predictor. The motion prediction $\Delta\hat{\boldsymbol{p}}$ is transformed from the predictor coordinate frame $\hat{T}$ to the vehicle coordinate frame T. The tracking error is measured in the reference coordinate frame R

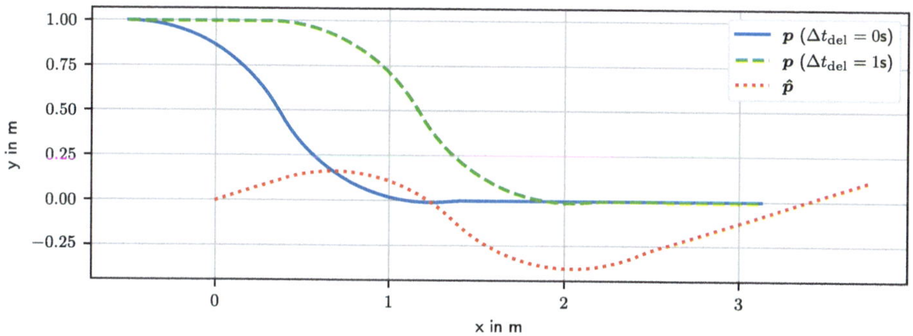

Fig. 7.8 Paths of the Dubins robust controller with dead-time compensation. The simulation plant is a kinematic singletrack model first without dead time and then a dead time of 1 s. $\hat{\boldsymbol{p}}$ is the path of the predictor plant for compensating the dead time

model path matches exactly the shape of the real path. Consequently, applying the rotation $\boldsymbol{R}(\psi_i - \hat{\psi}_{i-k})$, the prediction $\hat{\boldsymbol{p}}_i - \hat{\boldsymbol{p}}_{i-k}$ can be applied to the real position $\boldsymbol{p}_i$ to compute an (in this case) exact prediction of the real position $\boldsymbol{p}_{i+k}$.

7.3.1 Implementation and Stability

The dead-time compensation requires implementation of the plant model (7.8) and the state prediction (7.13). For the latter, the k previous estimated states $\hat{\boldsymbol{x}}$ are stored in a queue $\left[\hat{\boldsymbol{x}}_{i-k}\ \hat{\boldsymbol{x}}_{i-k+1} \cdots \hat{\boldsymbol{x}}_i\right]^T$. As discussed before, the state $\hat{\boldsymbol{x}}$ may diverge unboundedly.

To prevent numeric difficulties, in each time step i, we shift the predicted path to the coordinate center $\mathbf{0}$. Furthermore, we project the orientation $\hat{\psi}$ onto the interval $[0, 2\pi)$. The prediction model state queue contains the k last shifted position estimates and the k last orientation estimates:

$$\hat{\boldsymbol{x}}_{\mathrm{q}} = \begin{bmatrix} \hat{\boldsymbol{p}}_{\mathrm{q},i,-k} & \hat{\boldsymbol{p}}_{\mathrm{q},i,-k+1} & \cdots & \hat{\boldsymbol{p}}_{\mathrm{q},i} \\ \hat{\psi}_{i-k} & \hat{\psi}_{i-k+1} & \cdots & \hat{\psi}_i \end{bmatrix}^T \in \mathbb{R}^{k\times 3} \tag{7.14}$$

This queue describes the state increment from time step $i - k$ to each subsequent time step. The prediction model state queue $\hat{\boldsymbol{x}}_{\mathrm{q}}$ is computed as follows:

$$\hat{\boldsymbol{x}}_{\mathrm{q},i+1} = \begin{bmatrix} 0\;1\;0 \cdots & 0 \\ 0\;0\;1\;\;0 & \cdots 0 \\ & \vdots \\ 0 & \cdots \quad 1 \end{bmatrix} \cdot \hat{\boldsymbol{x}}_{\mathrm{qs},i} + \begin{bmatrix} 0 \\ \vdots \\ 0 \\ f_{\mathrm{pm}}(\hat{\psi}_i, \delta_i) \end{bmatrix} \tag{7.15a}$$

$$f_{\mathrm{pm}} = \begin{bmatrix} \boldsymbol{R}(\hat{\psi}_i) \cdot \frac{v}{\dot{\psi}(\delta_i)} \cdot \begin{bmatrix} \sin(\dot{\psi}(\delta_i) \cdot \Delta t) \\ 1 - \cos(\dot{\psi}(\delta_i) \cdot \Delta t) \end{bmatrix} \\ \dot{\psi}(\delta_i) \cdot \Delta t \end{bmatrix} \tag{7.15b}$$

$$\hat{\boldsymbol{p}}_{\mathrm{qs},i} = \hat{\boldsymbol{p}}_{\mathrm{q},i} - \hat{\boldsymbol{p}}_{\mathrm{q},i,-k} \cdot \mathbf{1}_{2\times k} \tag{7.15c}$$

$$\hat{\psi}_{\mathrm{qs},i} = \hat{\psi}_i \mod 2\pi \tag{7.15d}$$

where f_{pm} is the state increment of the plant model in equation (7.8) with $\hat{\boldsymbol{x}}_{i+1} = \hat{\boldsymbol{x}}_i + f_{\mathrm{pm}}(\hat{\psi}_i, \delta_i)$ and $\hat{\boldsymbol{x}}_{\mathrm{qs}}$ is the shifted state, containing positions $\hat{\boldsymbol{p}}_{\mathrm{qs}}$ and orientations $\hat{\psi}_{\mathrm{qs}}$. $\mathbf{1}_{k\times 1}$ is a $k \times 1$ matrix with value 1 in every entry.

With (7.15c), the predicted shifted position $\hat{\boldsymbol{p}}_{\mathrm{qs},i,-k}$ is $\mathbf{0}$ by definition. The positions $\hat{\boldsymbol{p}}_{\mathrm{q},i}$ will remain in a region around $\mathbf{0}$ that is bounded by the finite dead time k. Thus internal stability is obtained and the prediction model is BIBO stable.

The state prediction is computed according to (7.13) and using $\hat{\boldsymbol{p}}_{\mathrm{qs},i,-k} = \mathbf{0}$:

$$\boldsymbol{x}_{i+k} \doteq \boldsymbol{x}_i + \begin{bmatrix} \boldsymbol{R}(\psi_i - \hat{\psi}_{i-k}) \\ 1 \end{bmatrix}^T \cdot \begin{bmatrix} \hat{\boldsymbol{p}}_{\mathrm{qs},i} \\ \hat{\psi}_i - \hat{\psi}_{i-k} \end{bmatrix} \tag{7.16}$$

7.4 Evaluation

To evaluate the effect of the dead-time compensation, we simulate a path tracking scenario with the mentioned controllers and the presented approach. The vehicle is modelled by a kinetic single track model which differs from the prediction model due to the tire slip. Additionally we induce errors in the dead-time and the vehicle parameters.

7.4.1 Path Tracking Scenario

The plant system is a kinetic single track model implementing a Pacejka tire model [8] and load transfer. It has a minimum turning radius of 3.8 m. The input and output delay is composed of a steering system and a constant dead time modelling delays due to computation time and data transmission. The steering system is modelled as a third order low pass filter with a step response as shown in Fig. 7.9. The constant dead time is set to a value of 0.27 s. Note that the dynamics of the kinetic vehicle model also generate additional delay effects.

The reference path is composed of a right-hand-bend with radius of 27 m and a left-hand-bend with an increasing radius of 27–100 m and the vehicle is driving at a speed of $v = 11.1$ m/s.

7.4.2 Evaluation Results

We evaluate the path tracking scenario with the three controllers previously introduced. For each controller, the dead-time compensation is configured to compensate a dead time of

- $\Delta t = 0.2$ s: The dead time is underestimated
- $\Delta t = 0.4$ s: Considering the low pass filter in combination with the steering rate, and the additional dead time in the plant system, this value approximately conforms with an effective dead time in the controlled system
- $\Delta t = 0.5$ s: The dead time is overestimated.

The results are shown in Figs. 7.10 and 7.11. In all three controllers the effect of the dead-time compensation is qualitatively similar: the control performance with correctly parameterized

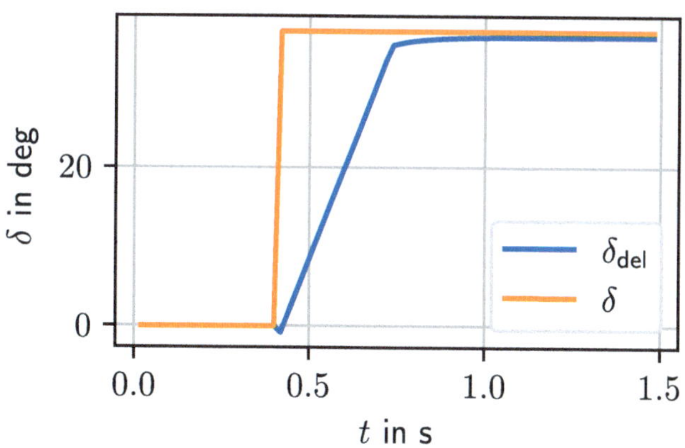

Fig. 7.9 Step response of the steering model which is implemented as a third order low pass. The steering model is implemented with $\Delta t = 0.001$ s to resemble the steering system which is continuous in time

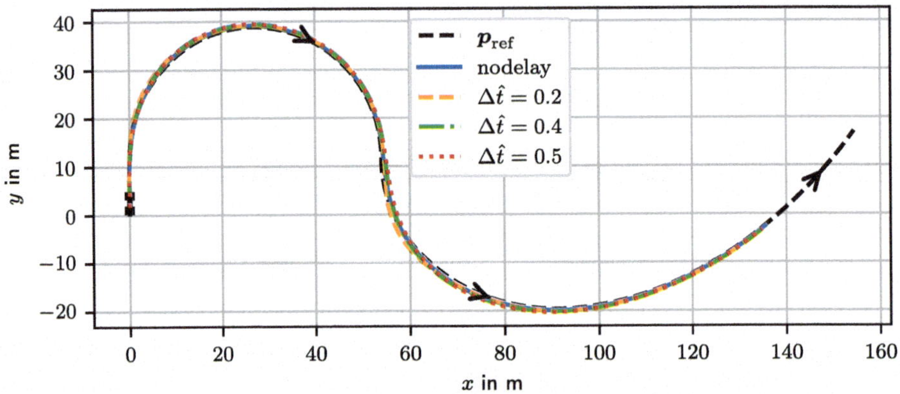

(a) Paths driven by the vehicle.

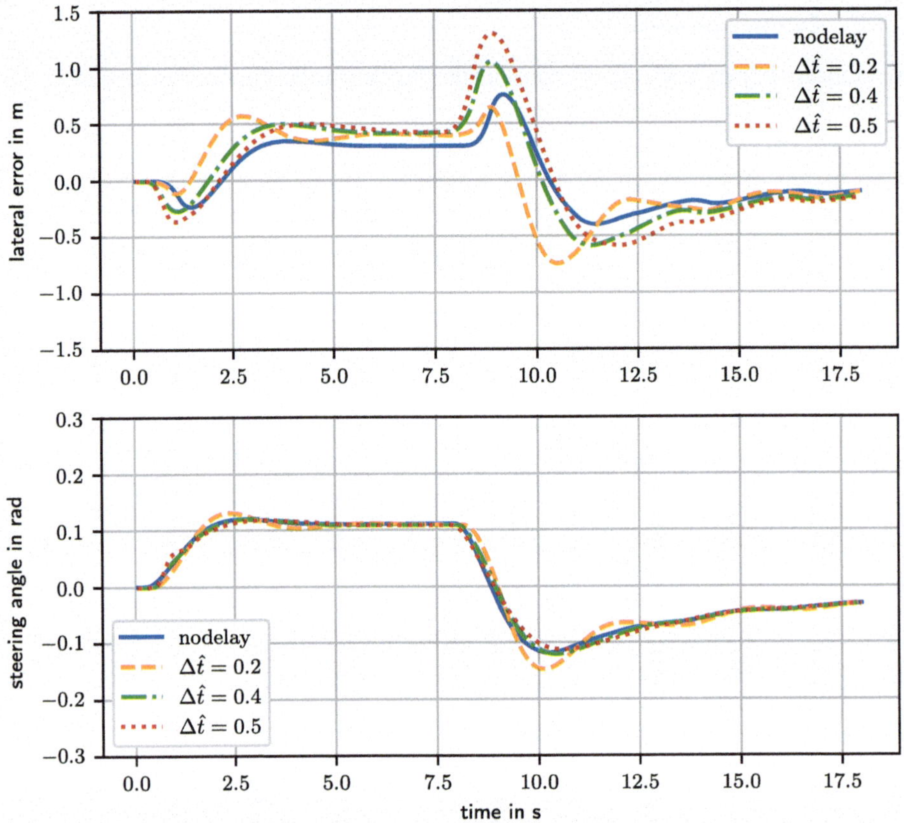

(b) Tracking error and steering angle during the scenario.

Fig. 7.10 Simulation of the Pure pursuit controller. The plant model has first no dead time (blue) and then a dead time of 0.4 s. The dead-time compensation is parameterized with different values for $\Delta\hat{t}$

dead-time compensation is worse than the control performance without dead time in the plant. However, the control performance with dead-time compensation is far better than without compensation and also better than with underestimation of the dead time.

The dead-time compensation method heavily depends on a good approximation of the plant system. Therefore, we evaluate how the control performance changes when the vehicle parameters are not exactly known. More specifically, we set the vehicle wheelbase $\hat{l}$ in the predictor plant to

$$\hat{l} = 0.9l$$

Moreover, in the derivation of the dead-time compensator we assumed constant velocity of the vehicle. To show the effect of a varying velocity, we add an acceleration from 5.5 m/s to 12.5 m/s at the beginning of the scenario. The results are shown in Fig. 7.12. The tracking

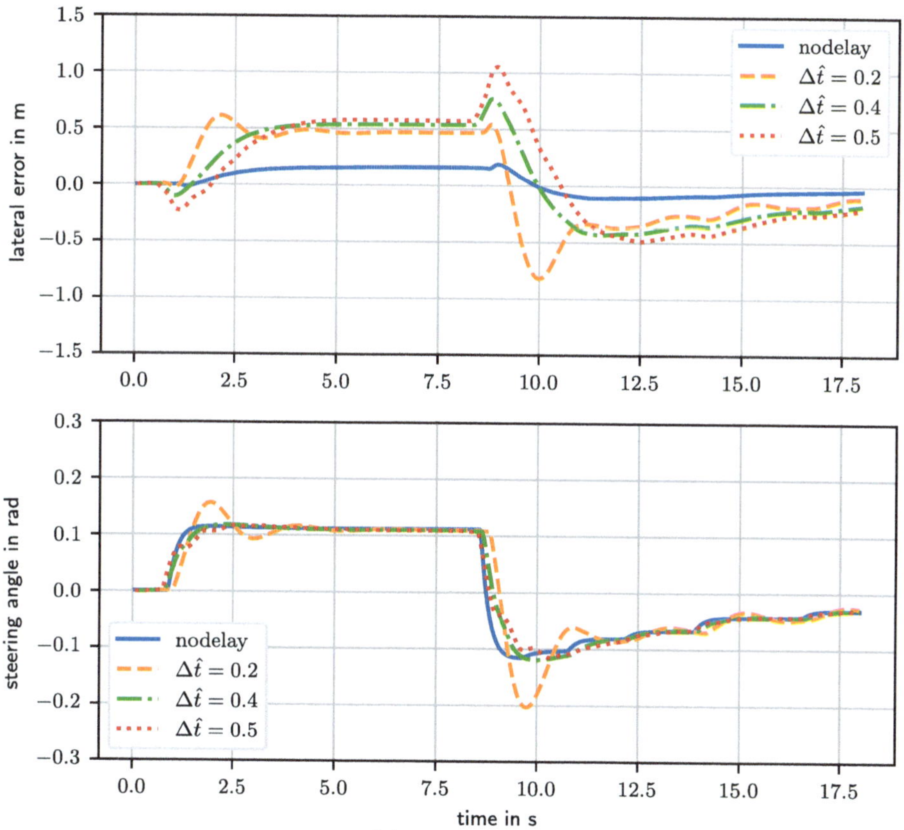

(a) Stanley

Fig. 7.11 Simulation of two controllers in the same scenario as in Fig. 7.10

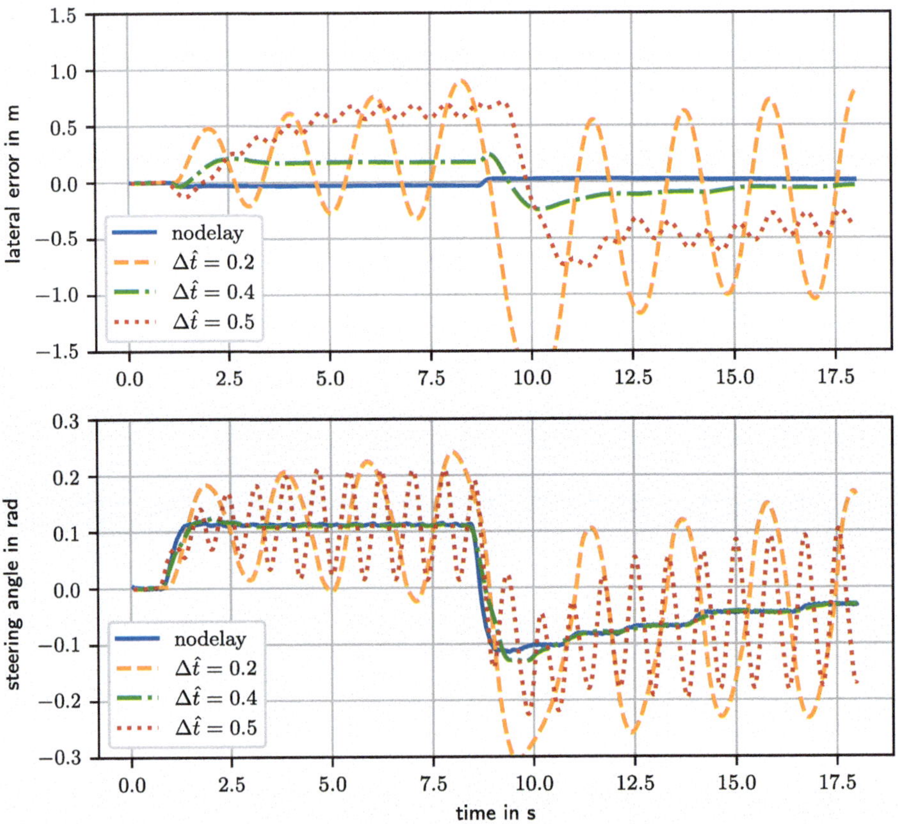

(b) Dubins robust control

Fig. 7.11 (continued)

error of the correctly parameterized controller is larger than in the previous simulation, due to the acceleration. This tracking error results from under-steering effects due to velocity dependent side slip.

The control error only slightly increases with the parameter uncertainty. When the wheelbase is underestimated, the predicted path has a larger curvature than the real path, resulting in a positive lateral error (with an overestimation of the wheelbase, the error would compensate the error caused by side slip, resulting in a smaller total error). Beside the change in tracking error, the deviation of the wheelbase estimate leads to oscillations in the steering angle at high velocity.

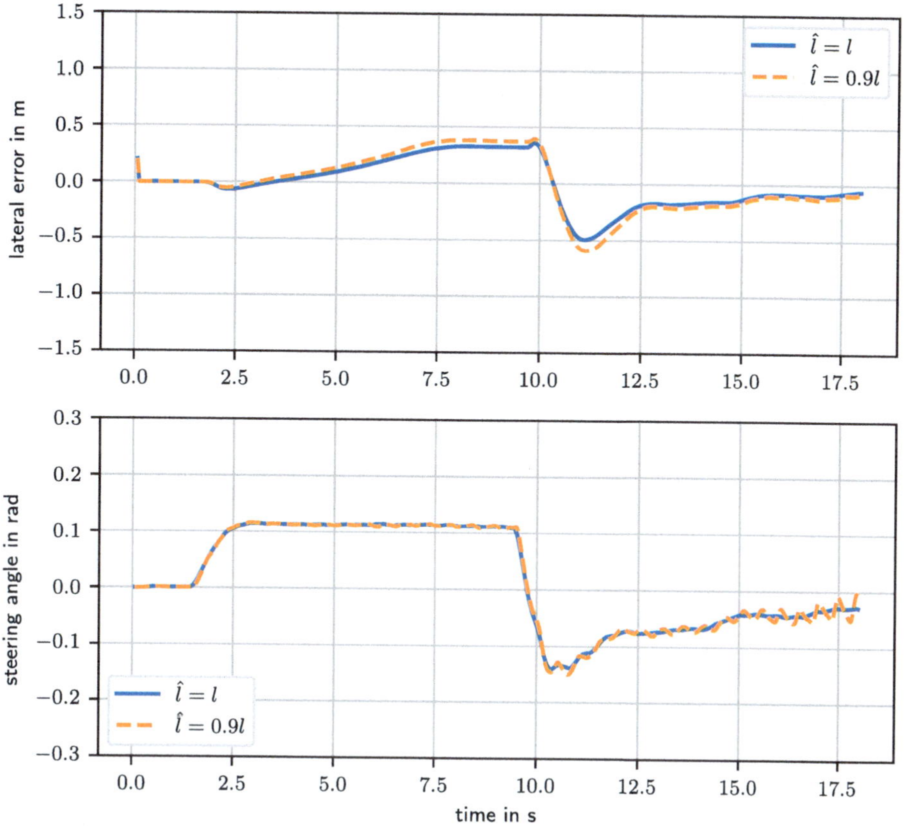

Fig. 7.12 Simulation of the Dubins robust controller. The scenario is the same as in Fig. 7.10 except that we vary the wheelbase $\hat{l}$ in the predictor plant and the velocity v is not constant

7.5 Conclusion

In this chapter, we addressed the significant challenge posed by dead time in the context of vehicle control, particularly in the application of classical path tracking controllers for autonomous vehicles. The simulation results consistently demonstrated the poor performance of various controllers when confronted with dead time, underscoring the pressing need for an effective compensation method.

We proposed a dead-time compensation method with a structure similar to a Smith predictor but with a state prediction that incorporates the specific nonlinearities of the kinematic vehicle model. This design makes a convergence of the prediction model not necessary and yet guarantees a bounded feedback correction. In simulation the method exhibited stability and significantly improved control performance also in presence of disturbances and parameter uncertainties. The effective dead-time compensation method shows promise as a practical solution to reduce the effects of dead time in fast-paced environments where quick path planning and tracking is crucial.

References

1. Coulter, C.: Implementation of the pure pursuit path tracking algorithm. Technical report, Carnegie Mellon University, Pittsburgh, PA (1992)
2. Guechi, E.-H., Lauber, J., Dambrine, M., Klančar, G., Blažič, S., Klančar, G., Blažič, S.: PDC control design for non-holonomic wheeled mobile robots with delayed outputs. Intell. Robot Syst. **60**, 395–414 (2010)
3. Hanus, R., Kinnaert, M., Henrotte, J.L.: Conditioning technique, a general anti-windup and bumpless transfer method. Automatica **23**(6), 729–739 (1987)
4. Hoffmann, G.M., Tomlin, C.J., Montemerlo, M., Thrun, S.: Autonomous automobile trajectory tracking for off-road driving: controller design, experimental validation and racing. In: Proceedings of the American Control Conference, pp. 2296–2301 (2007)
5. Hölzl, S.L.: The case for the Smith-Åström predictor. J. Process Control **128**, 103026 (2023)
6. Normey-Rico, J.E., Camacho, E.F.: Control of Dead-Time Processes. Springer International Publishing (2007)
7. Normey-Rico, J.E., Goh Mez-Ortega, J., Camacho, E.F.: A smith-predictor-based generalised predictive controller for mobile robot path-tracking. Control Eng. Pract. **7**, 729–740 (1999)
8. Pacejka, H.B., Bakker, E.: The magic formula tyre model. Vehicle Syst. Dyn. **21**(sup001), 1–18 (1992)
9. Tieber, K., Rumetshofer, J., Stolz, M., Watzenig, D.: Sub-optimal and robust path tracking: a geometric approach. In: 2021 IEEE/RSJ International Conference on Intelligent Robots and Systems (IROS), pp. 8381–8387 (2021)

The manufacturer's authorised representative in the EU is Springer Nature Customer Service Centre GmbH, Europaplatz 3, 69115 Heidelberg, Germany. If you have any concerns regarding our products, please contact ProductSafety@springernature.com

Printed and bound by CPI Group (UK) Ltd, Croydon, CR0 4YY
21/09/2026
02219994-0003